The Acquisitors

TOO TITANIC TO LET SINK

John Winslow

ISBN: 1439269068
ISBN-13: 9781439269060
Library of Congress Control Number: 2009913512

Also by John Winslow

The Accurst Tower

Contents

Contents

Tables

Preface

"Too big to fail." Does that mean big firms rescued during the Great Recession of 2009 never failed? With whiplash irony those four words extol the very quality that brought them down.

The big firms became too big by acquiring other corporations. They're so big we can't live without them, we're told.

We've forgotten that other corporate acquisitors made the same spiel long before the Great Recession began. They convinced us their bigness was good for us, so we've printed meltdown billions to save bigness.

The opportunity to escape bigness was served us on a silver platter. Had we not thrown the platter to the floor, the meltdown could not have happened. Period.

Published in its original form by Indiana University Press, *The Acquisitors: Too Titanic to Let Sink* looks at shady pitfalls from bigness and our spurned opportunity to correct them. It draws from the 1970s investigation of those pitfalls by the Judiciary Committee of the U.S. House of Representatives, which I served as a counsel.

Acquisitors collapse from bigness while they seek freedom from competition's controlling hand. We're told government regulation will control bigness better than the competition they evaded. This book tests that claim with past failures of regulation.

"No need to restore our broken waterworks plant," the villagers said. "Easier to *invent* a pill for diphtheria." *The Acquisitors* asks: Can we ever *invent* rules to restore the "invisible hand" of true market rivalry stifled by bigness? Why do we hesitate to break up acqusitors too titanic to let sink when our history since the break up of Standard Oil in 1911 is replete with restructuring?

"With all the corporate and regulatory horrors dredged up in this book and no effort by Congress to remedy them, perhaps Mr. Winslow will consider a sequel," Peter C. Ward, Bureau of Competition, Federal Trade Commission, said in *The Antitrust Bulletin* of *The Acquisitors* when first published. For a sequel, we have only to turn on news of the current horrors—horrors born after the day we threw the platter to the floor.

JFW
Washington, D. C
2010

PART I

✧ ✧ ✧

The Forces of Concentration

CHAPTER 1

Icebergs Unseen

Built of cheapest metal, the *Titanic*'s hull slid from its scaffolding into Belfast's tranquil waters. Waving to cheering crowds from the high platform, the president of White Star Lines turned to his chief architect to congratulate him for cutting costs. *What's wrong with cheap metal?* he must have mused. *I don't see any icebergs.*

But the liner's passengers paid for protection from *unseen* dangers. Americans, likewise, are shortchanged of protection from unseen dangers—dangers invented by corporations *too titanic to let sink.* We fully paid for a competitive market structure over the past century that would have protected us from unseen subprime mortgage crises if only we had maintained it. When did we squander it? When did we strike the iceberg?

August 26, 1971. On that forgotten day, Richard W. McLaren, assistant attorney general for antitrust, announced that the Justice Department had abandoned its suit to prevent International Telephone & Telegraph's takeover of Hartford Fire Insurance Co.—then the largest acquisition of all time. He gave no plausible reason.

If the Justice Department had not thrown that lawsuit to the winds, we would have steered clear of the present catastrophe: our printing hundreds of billions of dollars to rescue corporations too titanic to let sink. That's because the long-heralded lawsuit was to establish once and for all that none of the two hundred largest U.S. corporations could acquire an enterprise known to be a leader in any field. Obviously, Citigroup, American International Group, Wachovia, Bank of America, JPMorgan Chase, et al. could not have acquired other companies to become too titanic to let sink if that limit had prevailed.

Abandonment of the ITT-Hartford suit set off the chain reaction of bank takeovers (see p. 32). Can anyone think that over a thousand banks *all* would have invested in subprime mortgages before Bank of America took them over, shackled that investment to them, and sank them all? Would the shareholders of those thousand companies on their own have approved the salaries their acquisitor paid—amounts unknown to history?

The ITT-Hartford suit to prevent any major takeover by a major acquirer was to add better metal for protecting our right to be free of combinations that "restrain commerce." The Sherman Act (26 Stat. 209) intended to give us that right in 1890 (see p. 21). Congress reasoned that natural striving among competitors ordered by Adam Smith's "invisible hand"—not artificial regulation—would best ensure honest prices and square deals in the marketplace. But after the government scrapped the lawsuit, it proceeded to scrap *existing* protective metal.

ITT and Hartford did not function in the same industry. They could not form a monopoly and set prices. So why did scholarly Mr. McLaren work to keep them apart? He was not one to read other people's mail. But the acquisitors' private letters and memos showed him that megamergers not even forming monopolies can nonetheless skew the market system so that under it lurk age old icebergs: back-scratching trades between megabuyers and megasellers to undercut competitors, an employer's orders to its amassed employees to buy only its products, trades where seller is buyer and buyer, seller (heads I win, tails you lose), government rules that destroy market rivalry under the guise of protecting it, conflicts of interest, gentlemen's agreements not to compete, acquisitors' boasts of overwhelming the free market through sheer force of size (see chapter 14).

As Assistant Attorney General McLaren read, he saw man-made rules melting before ancient perils the letters plotted. First published by Indiana University Press in 1974, entitled *Conglomerates Unlimited: Failure of Regulation*, this book explores why the letters convinced him that natural rivalry is the true safeguard.

Audacity Works

The more audacious the takeover scheme, the more easily it overwhelms government safeguards, the letters showed Mr. McLaren. He read how the Penn Central Railroad with its mass of operations time and again overwhelmed transportation and Securities & Exchange Commission rules—rules meant to halt its colossal takeovers and switch it off Suicide Route. *But it's too titanic to let sink,* the Federal Reserve Board chairman believed (see chapter 11).

Mr. McLaren read of Litton Industries, acquisitor of 119 corporations over the past nine years, contracting to build the navy's largest vessels in a non-existent shipyard, then billing the navy for those vessels-never-built to gain more money to accelerate its rate of takeovers (see chapter 10).

The antitrust chief saw the advertisements (thousands) sent by Sanford Weill and his firm to debt-straddled companies telling them how to take over the treasuries of old-line insurance companies. After those takeovers, Mr. Weill himself became CEO of the insurance company that acquired Citicorp to form the super titanic, Citigroup. It received over $43 billion in bailout funds (see chapter 6).

He read memorandums of Chase Manhattan Bank telling of its concert with a giant acquisitor for rigging the stock market. He saw the concert perforating government safeguards (see chapter 4).

Icebergs change shape but never melt, Mr. McLaren concluded. *Audacity never fails.*

Immune to Madoff

Chase Manhattan acquired J.P. Morgan Co., to form the super titanic JPMorgan Chase in 2000—another type of combination the ITT-Hartford suit was to prevent. It, too, received $43 billion in bailout funds.

JPMorgan Chase invested a quarter billion dollars in Bernie Madoff securities in 2003. It lost not a cent. That's because Madoff deposited billions in its depository Chase Bank (formerly independent). From inside information gleaned from fluctuations of that deposit, JPMorgan Chase learned that Madoff was about to fold and sold out in time. The combined bank did not share that data with its customers for whom it had also bought Madoff securities. They lost everything. If they, too, had sold earlier the securities would have declined faster in value. JPMorgan Chase might not have known to get out in time.[1] The Glass-Steagall Act of 1935 (32 Stat. 160) forbade an investment bank from taking control of a depository bank for exactly that reason: an investor might learn inside information to use against other investors.[2]

But CitiCorp and Chase Manhattan told Congress the act should not stand in the way of audacious takeovers. Thus, it was part of the existing metal scrapped along with the ITT-Hartford suit.

The scrapping of the Glass-Steagall Act opened the way for investment banker Goldman Sachs to change into a commercial bank holding company in order to take bailout billions along with CitiGroup and JPMorgan Chase. Goldman Sachs made "huge profits" by selling risky securities to its customers and then shorting those same securities, i. e., betting against its own customers. "That's a most cynical use of credit information," more than one financial expert says.[3] Now that Goldman Sachs is free to accept customers' deposits—as it could not before the act's repeal—it should enjoy vastly greater resources for selling securities it later shorts.

It does already. "Goldman Sachs' and other banks' role in masking Greece's mounting debts may be pushing the nation closer to the brink of financial ruin," *The New York Times* reported February 25, 2010. "If Greece reneges on its debt, these credit-default swaps [purchased by the banks] will enable them to profit..." by selling short. Understandably, among them is commercial-investment bank JPMorgan Chase (see p. 71).

Tentacles Set Free

Even the protective metal that forbad competitors to combine, the Celler-Kefauver Amendment of 1950 (38 Stat. 730), has been largely scrapped. The voluminous list of insurance companies American Insurance Group swallowed after August 26, 1971— rushing to extend its "tentacales throughout the economic system" before its demise—shows regular breach of that hard-won safeguard. Ironically, AIG had to transfer billions of its own life-support funds to those companies that prospered before AIG took them over—not just other insurance companies but utilities, a national telephone company, the world's largest purchaser-lessor of aircraft (see p. 299).

Scrapping of that metal does not worry Jamie Dimon, CEO of JPMorgan Chase, employing "more than 220,000 people, serving over 100 million customers, lending hundreds of millions of dollars each day, and operating in nearly 100 countries." Too-titanics-to-let-sink, entangled in the economy though they are, no longer pose danger because regulators can simply stack new artificial rules onto piles of failed rules, he says. CEO Dimon envisions no more injury to the public from their sinking. We must give them free reign to continue to acquire other corporations and certainly not break them into parts, he warns.[4]

But we tow those acquisitors to shore because they grew too titanic after we scrapped the ITT-Hartford suit. How can we avoid devastating towing bills *without* breaking them into parts?

Antitrust Chief McLaren sought to spare us those bills by keeping the parts separated in the first place. To see the perils he saw then—and knew would haunt us now—is the purpose of this book.

CHAPTER 2

The Offensive Intensifies

My power would fall were I not to support it by new expansion.
Conquest has made me what I am, and conquest shall sustain me.
— Napoleon

A morning ray of Nixon idealism was appointment of noted antitrust litigator Richard W. McLaren to be assistant attorney general for antitrust. In those early days, Mr. McLaren feared that giant acquisitors were sustaining their expansion not with better service but, as Napoleon, by conquering more prizes. Brow furrowed, the antitrust chief examined the charts and detected a pattern. The hundred largest U.S. industrial corporations in 1970 controlled a greater share of the nation's manufacturing facilities than did the two hundred largest in 1950. The two hundred largest industrial corporations in 1970 in turn held more than 60% of all manufacturing plants and equipment—a greater share than the one thousand largest companies owned three decades before.[1] The vast increase of assets those corporations control comes not from their building new plants and equipment but rather from their taking over other companies.

> Explosive growth by acquisition by our largest corporations presents the danger that the American economy will be dominated by virtually self-contained economic domains.

Growth by these vast corporate structures presages imposition of cartel-like structures throughout American business…

the report by the staff of the House Judiciary Committee, *Investigation
of Conglomerate Corporations*, p. 439 (1971), warned.

It pointed to the twenty-five acquisitors that had expanded
most greatly into multifarious industries by purchasing other corporations, which *Fortune,* February 1969, listed. In eight years the
twenty-five had acquired 695 companies costing altogether over $20
billion (in 1960s dollars). Those billions of assets, rather than spread
among 695 independent owners as before, were concentrated in
twenty-five corporate headquarters. The concentration meant, he
knew, a disastrous scheme could spread among those corporations
with a chance 27.8 times smaller (the 695 independent minds of
the former owners divided by 25) of being detected. Little did he
know that the viral subprime mortgage scheme would spread with
a chance a thousand times smaller of being detected—after a single
bank, Bank of America, took over a thousand banks and bestowed
the scheme on them.

Not surprisingly, the *Fortune* list shows that the twenty-five acquirers' assets rose astronomically over other corporations' assets
during the decade. Obviously, their rise came merely from shifting
plants, offices, and machinery from the former owners to themselves—not from creation of new facilities that provide for industrial expansion, greater production, and increased employment.

Mr. McLaren saw that between 1948 and 1952, the cost of acquiring other corporations was less than 3% of the total expenditure on new capital investment. By 1968 that figure had risen to
55%.[2] Thus, U. S. corporations were spending more and more to
purchase existing facilities rather than to invest in new facilities. The

assistant attorney general for antitrust saw a threat to the economic system: without that investment the economy does not grow. And without increased production capacity, rising consumer demand results in rising inflation.

Baffled

The pretext acquisitors give for binding enterprises into a shrinking number of corporate giants is the advantage they claim to enjoy from rushing in many directions at the same time. The present phenomenon of consolidation would baffle masters of earlier eras of notorious corporate concentration, such as John D. Rockefeller, Andrew Carnegie, and J. P. Morgan. For they formed their industrial trusts and combinations each to monopolize a distinct field of industry. Standard Oil Company would scarcely have bothered to acquire thirty-seven other companies by 1911 had those companies functioned outside Standard's target industry for monopolization. The same singleness of purpose—elimination of competitors— gave rise to other monolithic combinations such as the steel and railroad trusts. The American Tobacco Company's acquisition of its competitors was so devoid of any purpose other than the elimination of rivals that it promptly scrapped their plants, warehouses, and equipment upon purchase. Present-day acquisitors hold the assets of their acquired companies in far higher esteem.

By monopolizing an entire industry, and hence expelling competition, the dominant company can set prices and standards of performance at its whim. Congress sought to curtail those abuses by enacting legislation beginning in 1890 and culminating with the 1950 Celler-Kefauver Amendment to the Clayton Antitrust Act. Its object was to prevent combinations of corporations engaged in the same industry.

They All Sink

Why the great merger era of the mid-twentieth century in-volved combinations of enterprises in dissimilar enterprises was, then, no mystery. The question, rather, was why the amalgamation of companies in unrelated fields occurred at all. Specifically, whether the acclaimed rewards of diversification resulting from taking over diverse corporations was the true cause. Roy Ash, former president of Litton Industries (in 1967 and 1968 the acquirer of companies at the rate of almost two a month), likened that reward to an ocean fleet's agility:

> I like the analogy of the aircraft carrier and the destroyers. Think of a larger, centralized company as an aircraft car-rier. If a man falls overboard, you'd have to turn the whole ship around to go back after him. Litton is more like a fleet of smaller destroyers. If a man falls overboard, only one destroyer has to turn back to pick him up. The others can continue unimpeded with their mission.[3]

But if the small destroyers are consolidated into one titanic and it hits an iceberg, they all sink. Mr. Ash wasn't worried.

In 1969 and 1970, ten investigations of national scope were at-tempting to find less watery explanations for the wave of corporate consolidations. The Senate Antitrust and Monopoly Subcommittee, the Federal Trade Commission, the Federal Communications Commissions, the Securities and Exchange Commission, the New York Stock Exchange, the House Ways and Means Committee, the Department of Justice, the Interstate Commerce Commission, the Cabinet Committee on Price Stability, and the House Antitrust Subcommittee were all conducting separate inquiries.

Fatal Oblivion

"Do the consolidations threaten free flow of commerce?" the investigators asked. Their discoveries are kept under wraps, if not disowned.

The Investigation of Conglomerate Corporations, 1969-1970, prepared by the Antitrust Subcommittee of the House Judiciary Committee, relies on the internal documents acquisitors generated while they acquired. Only the committee staff, not the congressmen it served, issued the investigation report. Similarly, the Federal Trade Commission's *Economic Report on Corporate Mergers*, 1969, was issued as a report of the commission staff. The commission itself, created to protect the public from undue economic concentration, disclaimed the report with the introductory comment: "The Commission as a whole and the individual Commissioners neither necessarily endorse nor adopt the report or its recommendations."

Fatally ignored, the Interstate Commerce Commission staff report warned the commissioners more than a year in advance of then history's largest bankruptcy, the collapse of Penn Central.

The Federal Trade Commission staff prepared a second study, *Conglomerate Merger Performance—An Empirical Analysis of Nine Corporations*, January 3, 1973. It collected performance data of 348 acquired subsidiaries. That information, however, revealing whether specific acquiring managements enhance or hinder the U.S. economy, was deleted from the publicly distributed analysis. The Federal Trade Commission thus protected those "managements from their stockholders by keeping specific mistakes secret," according to one of the three staff economists.[4] "The commission destroyed all copies of the 'hot report' by putting them through a shredding machine."[5] But the most reticent of the antitrust investigators is the Federal Communications Commission. Beginning in 1970 it employed a staff to investigate mergers and acquisitions in the communications industry. By mid-1973, a year and a half after completion of the study, the commission had given no explanation for not having released a single word of its investigation findings. The public must be spared them.

Claims in Conflict

Acquisitors should welcome that abundant disclosure secured at public expense for arguing their claims that:

- Welding together diverse enterprises under a parent acquirer strengthens the parent by diversifying its activities.
- The parent acquirer's management of those welded-together companies is superior to the formerly independent management of each, which obviously operated the company successfully. Why else would it be purchased?
- The acquirer injects stamina into its acquired subsidiaries—financial and technological.

Likewise, the examination should test the charges that:

1. The word "diversification" by its mere resonance stifles common sense. Investors mistake the shifting of wealth (assets and earnings of acquired companies) for the creation of wealth (after it is added to the acquisitors' coffers). Only when the acquisitions cease do they see that acquisitors' inflated earnings reports come not from heightened performance but from acquiring other companies' assets. Acquisitors falsely claim success for diversification by

simply adding to their own income the earnings of acquired companies.

2. Excuses given for acquisition often are objectives that managements could patently better accomplish without merger. Acquisitors even promise shareholders of target companies, in return for their consent to be bought out, payment that only the assets they surrender can produce.

3. Many acquisitors must acquire to survive. Shareholders of healthy corporations, nevertheless, rush to shackle their assets to moribund corporations that exist only from the strength of the companies they acquire.

4. The more prosperous the enterprise, the greater its chances to be taken over. More than one one-third of the thousand largest American manufacturing corporations of 1950 had been acquired by other companies by 1970.[6] In previous decades, less prosperous companies were the object of merger. The purpose, theoretically, has been to transfer assets, usually of smaller companies, to a more successful management. The current merger era shatters that tradition. Ironically, the former assistant attorney general for antitrust, Professor Donald Turner, told the Senate Antitrust and Monopoly Subcommittee that good management is a company's best protection against being acquired.[7] As prevention against acquisition, he prescribes the exact quality that drastically increases a company's chances of becoming an acquisition target.

5. Acquisitors are loath to reveal whether their acquired companies benefit from merger. Data subpoenaed from them, however, indicate that performance of acquired companies declines under the amalgamated control.

6. The undisclosed declines of the acquired parts draw the acquisitor into a vicious cycle. It can show increased earnings and claim improved performance only by acquiring still more companies and adding their earnings to the total. Automatic Sprinkler, for one, was in that cycle when the decreasing earnings of its acquisitions were insufficient to purchase more companies. Then, for the first time, it could not inflate acquired earnings with more acquired earnings. Profits dropped from $1.43 to ten cents a share.[8]

7. Another device for claiming a merger's benefit is a change in accounting methods. After acquiring John Morrell & Co., AMK Corporation changed Morrell's depreciation and inventory accounting and thereby reported a 74-cent-per-share profit increase although the profits of the acquired company had actually declined. Gulf & Western Industries likewise, by shifting accounting methods, showed an earnings increase of $1.6 million for New Jersey Zinc and a multimillion-dollar earnings increase for Paramount Pictures.[9] Those ingenious reports are from the accounting gimmickry sanctioned by federal regulatory agencies and the accounting profession.

8. The acquisitors, masters of finance rather than operations, secure those paper profits from acquired companies, inflate the market value of their parent-company stock,

and thereby secure more resources for acquiring more companies.

9. Corporate acquirers resort to explanations of declining profits that absolve themselves from blame—explanations their private documents specifically refute.

10. Unable to rely on the competitive quality of their products and services, acquisitors purchase other companies to manipulate customers through reciprocal and tie-in sales. Their practices belie their disclaimers of anticompetitive intent.

11. The acquisitions deprive investors of insight into corporate performance. The Securities and Exchange Commission exempts acquired corporations from disclosure requirements after they lose their independence. Ingalls Shipbuilding Corp., for one, is not required as a Litton Industries subsidiary to reveal performance data required of it when independent. Thus, the public did not learn of the failure of the most colossal shipbuilding contract of all time until it received the bill for Ingalls' cost overruns. A fifteen-volume claims "summation" to the navy is only the first chapter of that bill.

12. The spiral of industrial concentration could not flourish if the regulatory agencies performed merely their elementary duties. Government administrators, however, reject the findings of their own staffs in order to give special dispensations to permit the largest takeovers. Frequently, the

larger the illegal transaction, the greater the chance of official acquiescence.

The Securities and Exchange Commission refuses to divulge which specific investment companies escape the Investment Company Act's provisions for investors' security. The commission quickly approved, however, a spurious $2.3 million profit an acquisitor reported to the public in place of an actual $80 million annual loss.

Mr. McLaren did not charge that every acquisitor engages in all those practices. But he had seen enough. He convinced his nominal superior, Attorney General John N. Mitchell, that the Justice Department must use its legal might to stop the largest two hundred U. S. corporations from ever again purchasing another major corporation.

CHAPTER 3

Mounting the Counterattack

Attorney General John Mitchell had only praise for Adam Smith. The classical economist's "invisible hand" governed commerce as wisely as it did effortlessly, i. e., with minimum regulation, he told the Georgia Bar Association convention. The lawyers gave him a standing ovation that June afternoon, 1969. Mr. McLaren had written the speech for him. But the attorney general already doubted the acquisitors' claims. Communities throughout the country were losing their native enterprises and hence their economic destiny, he told the convention. Acquisitors were taking control away from the localities that had created the enterprises and transferring it to new headquarters in distant cities. Acquisitors were turning corporate birthplaces into "branch store communities":

> One of the great benefits of the open market place is the active participation and control by as many of our citizens as possible in their own economic well being—not just a small segment of our population in certain cities. An urban area should have a substantial influence over its local economy...And its consumers should have the opportunity to exercise local economic options in their choice of competing goods and services...
> We do not want our middle-sized and smaller cities to be merely "branch store" communities.[1]

California had lost the most corporate headquarters to other states—511—he said. New York had gained the most—1,732.

Cascade

Mr. Mitchell and his antitrust chief promised to assert their power with no holds barred "to meet all threats to competition" posed by the rising wave of acquisitions.[2] "The risk of losing some cases" was small cost for attempting to halt the merger surge in time, Mr. McLaren said. He warned would-be acquirers not to rely on the mild "merger guidelines" issued by his Democratic predecessor. He told the House Ways and Means Committee:

> I've warned businessmen and their lawyers that we may sue
> even though a merger appears to satisfy those guidelines…
> By no means am I opposed to amendatory legislation. But
> we cannot wait. We are willing to risk losing some cases
> to find out how far antitrust law will take us in halting this
> accelerated trend toward concentration by merger and, as
> I see it, the severe economic and social dislocations it will
> bring.[3]

"Antitrust law" is to protect market access and market rivalry. The economic, social, and political functions of that law are inseparable. For to deny market access is to foreclose means of livelihood, which in turn is to thwart creativity and independence:

> Competition is also desirable on principle and for its own
> sake, like political liberty and because political liberty is
> jeopardized if economic power drifts into relatively few

hands. Antitrust also performs the functions of keeping governing power in the hands of politically responsible persons. Power to exclude someone from trade, to regulate prices, to determine what shall be produced, is governing power, whether exercised by public officials or by private groups. In a democracy, such powers are entrusted only to elected representatives of the governed...Antitrust opposition to overwhelming Bigness serves still another purpose. Intellectual and artistic creativeness can be imperiled by the quality of sameness imposed on us when standards of thought are delivered into the hands of a few businessmen...[4]

Rivalry among competitors in the marketplace is a guarantor of the worth of their goods and services, and is thus a natural regulator of prices consumers pay. Enforcement of the few laws that protect competition, therefore, serves in the stead of vast artificial regulation.

Federal Reserve Board Chairman Arthur F. Burns was disdainful of Band-Aids. On March 22, 1973, he told the Senate Antitrust and Monopoly Subcommittee: "Selective controls on wages, and prices for a limited period will be helpful, but they are no substitute for vigorous competition."[5] Failure of competitive market structure leads to inflation, balance of payments crises, then to devaluation of the dollar, he warned.

He urged only fundamental reform: market structuring to protect competition. His successor at the Federal Reserve, Chairman Allan Greenspan, also disdained artificial control of commerce. He said banking had become too complicated for the government: "Regulators cannot do that job" (*The Washington Post,* December 21, 2009).

With Citigroup and Bank of America growing too titanic, however, Mr. Greenspan resorted to the quick fix of "reducing interest rates when speculative bubbles ran into trouble." His successor, Chairman Ben Bernanke, "set interest rates essentially at zero" (*The New York Times*, September 10, 2009).

But we're "always fighting the last war," when we simply respond to emergencies, said John McFall, chairman of the House of Commons Treasury Committee (*The Financial Times,* September 10, 2009). "The real problems are not the [immediate] causes of the crises of 2008 (banks) or 2001 (dotcom) or 1998 (Long-Term Capital Management) or 1989 (US Saving and Loans), but the enduring power of finance to be disruptive. Once new regulations are in place, the corporate sector is already involved in even more dangerous practices." We're dogs jumping through hoops.

Fixing the system "requires this business model to be broken up," Chairman Mc Fall said in 2009. But that non-competitive business model should not be allowed to coalesce in the first place, Attorney General Mitchell had avowed to the Georgia lawyers in 1970.

Besides "maximizing choice and utility for consumers," freedom of competition puts national resources to fullest use:

Not only consumers, but those who control the factors of production—labor, capital and entrepreneurial ability benefit when resources are permitted to move into the fields of greatest economic return. Competition induces such movement...It minimizes the necessity for direct Government intervention in the operation of business, whether by comprehensive regulation of the public utilities type or by informal and sporadic interference such as price guidelines and other ad hoc measures.[6]

Building the Dam

The Sherman Act of 1890 (26 Stat.209), outlaws "every...combination" or trust that inhibits market access and thus restrains commerce. It proscribes attempts to monopolize "any part" of interstate trade. "As a charter of freedom," Chief Justice Charles Hughes wrote in 1933 (*Appalachian Coals, Inc. v. United States,* 288 U. S. 344):

> The act has a generality and adaptability comparable to that found to be desirable in constitutional provisions. It does not go into detailed definitions which might either work injury to legitimate enterprise or through particularization defeat its purposes by providing loopholes for escape.

But the Supreme Court did not use the full force of the Sherman Act to halt concentrated control of industry. It did not outlaw "every...combination" that restrains trade—the act's stated purpose. In the Standard Oil decision of 1911 (221 U. S. 1) it invented the "Rule of Reason," allowing any industrial combination's restraint on trade that is somehow within reason. Justice John Harlan dissented:

> Congress...said that here should be no restraint of trade, *in any form*, and this court solemnly adjudged many years ago that Congress meant what it thus said in clear and explicit words of an act. But...now the Courts will allow such restraints of interstate commerce as are shown not to be unreasonable or undue.

The Senate Interstate Commerce Committee agreed with Justice Harlan. The "Rule of Reason" usurped power of congress: "It is inconceivable that in a country governed by a written Constitution and statute law the courts can be permitted to test each restraint of trade by the economic standard which the individual members of the court may happen to approve."[7]

By the time of the presidential election of 1912 the issue was clear-cut. Theodore Roosevelt's New Nationalism platform accepted trusts and combinations as a fact of life: they were to be held in check by vast government regulation, not competition. Woodrow Wilson's New Freedom policy, on the other hand, was to combat the persistent merger cycle by protecting the natural workings of the marketplace, i. e., by real enforcement of antitrust laws. Wilson warned of encrusted control centers:

> No country can afford to have its prosperity originated by a small controlling class. The treasury of America does not lie in the brains of the small body of men now in control of the great enterprises...It depends upon the inventions of unknown men, upon the originations of unknown men, upon the ambitions of unknown men. Every country is renewed out of the ranks of the unknown, not out of the ranks of the already famous and powerful in control.[8]

Wilson's optimism ostensibly won out. Reasoning that true market competition is the best protection against the creation of economic domains, Congress enacted the Clayton Act (38 Stat. 730) to prohibit specific transactions such as price discrimination, exclusive arrangements, tie-in sales contracts and corporate acquisitions if the "effect may be substantially to lessen competition or tend to create a monopoly." It is to protect market structure.

But antitrust schizophrenia won again. The courts "frustrated... Congressional design and as a result...[the Clayton Act] fell short of its intended purpose to stop in its incipiency undue concentration of economic power or monopoly."[9] Soon after World War II a new trend of corporate concentration arose that the restrained application of the Clayton Act failed to contain. Corporations engaged in the same industries were combining at an accelerating rate. Competition suffered because fewer independent rivals were left in the marketplace. Similarly, the rate of mergers between suppliers and their customers was increasing. Thus, producers were acquiring customers not because they produced the better product but because they controlled those customers. Suppliers lost sales to the growing number of acquired corporations regardless of the quality of their product.

Dam Break

Between 1948 and 1951, 62% of all corporate combinations involved such elimination of competitors.[10] Industrial concentration was at one of its highest levels, the House of Representatives reported in 1949, as it amended the Clayton Act "to prevent those acquisitions which substantially lessen competition or tend to create a monopoly."[11]

Congress thus acted "to restrain trends of concentration in their incipiency and well before they have attained such proportions as would justify a Sherman Act proceeding."[12] That act, the Celler-Kefauver Amendment of 1950, prohibits mergers of two corporations functioning in the same industry "horizontally," i. e., by producing similar products, or "vertically," i. e., by one corporation's supplying products to the other.

The amendment's success in restoring competition accelerated. Combinations of corporations engaged in the same industry had amounted to 48% of all mergers between 1952 and 1959. The ratio steadily fell, amounting to 22% between 1964 and 1967. In 1968, mergers of industrially related corporations amounted to only 9% of total mergers.[13]

But the inexorable current of corporate concentration simply changed courses. Because the Celler-Kefauver Amendment blocked the channel of combinations of enterprises operating in the same industry, the current merely shifted and overflowed with even greater force into the channel of mergers of corporations in unrelated industries. Assistant Attorney General McLaren aptly said in 1969: "Merger activity…has been redirected, not reduced."[14]

That industrial concentration had not abated during the two decades since the 1950 amendment to the Clayton Act was an understatement. The number of corporate mergers (or acquisitions) increased from 295 in 1953, to 844 in 1960, to 1,008 in 1965. Three years later the number of mergers annually had more than doubled, amounting to 2,442 in 1968. Assets acquired from other companies amounted to $953 million in 1953, $2,326 million in 1960, and $4,914 million in 1965. In three years that annual figure more than tripled, amounting to $15,200 million in 1968.[15]

The current wave of concentration threatens to restructure industry, Attorney General Mitchell continued. "The vitality of our free economy may be in danger," he told the lawyers in Atlanta:

> Concentration of this magnitude is likely to eliminate existing and potential competition. It increases the possibility for reciprocity and other forms of unfair buyer-seller leverage. It creates nation-wide marketing, managerial and financial structures whose enormous physical and psychological resources pose substantial barriers to smaller firms wishing to participate in a competitive market. And finally, super-concentration creates a "community of interest" that discourages competition among large

firms and establishes a tone in the marketplace for more and more mergers.[16]

"Fewer and fewer people" controlled the nation's industrial power. That's the "evil" the Sherman Act, the Clayton Act, and the Celler-Kefauver Amendment sought to eliminate, he said.

But he did not describe the grasp of merging competitors. For that reason, application of the antitrust laws to the phenomenon had not been proven, as it had been to earlier eras of industrial combination. Although the increase of mergers between dissimilar enterprises transferred greater and greater assets to fewer and fewer corporate owners, the transfer did not result in monopolization of fields of industry. The new owners were jacks of many trades but usually monopolizers of none.

A Stitch in Time

Nevertheless, Attorney General Mitchell promised, the Department of Justice would invoke the present antitrust laws, namely the Sherman Act and the amended Clayton Act, to restrain the new merger wave. He cited the Procter & Gamble decision of 1967 (386 U. S. 568) as authority that the government, in order to protect a market structure conducive to competition, need not prove specific anticompetitive behavior. The Supreme Court decided then:

> The Clayton Act was intended to arrest anti-competitive effects in their incipiency. The core question is whether a merger may substantially lessen competition, and necessarily requires a prediction of the merger's impact on competition, present and future...[Section 7 of Clayton Act] can deal only with probabilities, not certainties ...and there is

certainly no requirement that the anti-competitive power manifest itself in anti-competitive action before Section 7 can be called into play.

If enforcement of Section 7 turned on the existence of actual anti-competitive practices, the congressional policy of thwarting such practices in their incipiency would be frustrated.

He updated the merger guidelines (which Assistant Attorney General McLaren had said a few weeks earlier were outmoded) and warned that the Department of Justice "will oppose any merger by one of the top 200 manufacturing firms of any leading producer in any concentrated industry."[17]

Reciprocity

A primary threat giant mergers pose to competitive market structure, the attorney general and his antitrust chief explained, is reciprocal dealing. By that system, one corporation patronizes another in return for its patronage. A bank, for example, indulging in reciprocal practices, awards its contract for building construction to the concern that happens to be its biggest customer among the bidding contractors. Reciprocity of patronage thus renders futile any competition among the bidders.

If a supplier of goods has no occasion to purchase from one of its customers, of course no possibility of reciprocal sales between the two concerns exists. Reciprocity should not occur, therefore, between a newspaper publisher and its supplier of pulp wood because the pulp supplier probably spends less than a dollar a day on newspapers. Likewise, reciprocal sales should not be expected to arise between a specialty food processor and a cattle ranch from which the processor purchases beef, if the rancher has no need of the specialized processed food and never buys it from anyone. But if the pulp wood producer acquires the food processor, and the newspaper and cattle ranch likewise come under common management, the pulp producer can augment its sales to the newspaper by threatening to curtail its purchases of beef for its food processor subsidiary unless the publisher-rancher syndicate reciprocally purchases the pulp wood. Thus, a combination of companies in unrelated industries gives rise to reciprocity, from contexts in which such practices would be impossible but for the combination. The example is not hypothetical. The pulp wood producer is Rayonier Corporation, and the food producer is the nation's largest baker, Continental Baking, both owned by International Telephone & Telegraph (see chapter 8).

Through reciprocity a producer secures sales by the force of his purchasing power, not by improving his product or lowering his prices. Thus, widespread reciprocity might well necessitate government controls for the protection of consumers, for rivalry in the marketplace would cease to function as a natural insurer of quality and regulator of price. As the attorney general warned, if industrial assets are concentrated under fewer and larger control centers, "we may soon be in a position where demands for more government regulation could be called for."[18]

Prevent Future Rivals

The phenomenon of mergers between enterprises in unrelated industries poses a second threat to market structure because it can curtail future rivalry by reducing the number of entrants into an industry. Many corporations that enter new fields of industry by purchasing enterprises already functioning would otherwise enter by establishing new operations—and thus add to the number of competitors. As an illustration of how even the possibility of new competition in a market benefits consumers, the attorney general cited the natural gas industry. A certain retailer of gas, theretofore unrivaled, quickly lowered prices 25% on the mere expectation that a competitor was about to enter his market area. "Elimination of potential competition," Mr. Mitchell said, "tends to maintain the inflated price structure in the concentrated industry."

Gentlemen's Agreements

A third effect of industrial concentration on competitive market structure is blandly termed "community of interest." As more and more formerly independent corporations fall under the control of acquiring managements, the possibility looms larger of gentlemen's agreements among those managements to forebear from competition. *Studies by the Staff of the Cabinet Committee on Price Stability,* January 1969 (p. 83), observed that as more diversifying corporations encounter one another in more industries, the chances increase that competitive aggressiveness in one field will encounter retaliation in another:

> [Loss of independence by acquired corporations] create[s] the probability that a community of interest will develop among the leading actual or potential competitors in many industries. There is also a danger that industries once standing as separate arenas of competition will have their competitive zeal diluted as conglomerate firms operating across many industries meet the rivals in many markets.
>
> If they engage in aggressive acts in particular markets they may anticipate retaliatory measures elsewhere.

Attorney General Mitchell warned of "community of interest":

If the food subsidiary of corporation A aggressively competes with the food subsidiary of corporation B, then the electrical subsidiary of corporation B may start a price war with the electrical subsidiary of corporation A. Thus, it may be in both A's and B's interest to maintain the status quo and not to engage in the type of aggressive competition which we expect in a free marketplace.[19]

To evade a boomerang, a DuPont official urged forbearance toward Imperial Chemical Industries of Great Britain. The companies encountered each other in various endeavors. The official wrote:

It is not good business sense to attempt an expansion in certain directions if such an act is bound to result as a boomerang of retaliations. It has been DuPont's policy to follow such lines of common sense procedure.

This was done on the broad theory that cooperation is wiser than antagonism and that in the matter of detail the chances in the long run were that the boot was just as likely to be on the one leg as on the other.[20]

A "community of interests" calls, then, for "common sense procedure" rather than profit analysis.

Chain Reaction

The Department of Justice filed five suits to checkmate the exact dangers its two chiefs vividly forecast. The suits were based not on the conventional antitrust conception of monopolistic dangers arising from the combination of enterprises engaged in the same industry, but rather on the more untried theories of the dangers of reciprocal practices that Mr. Mitchell had fully described earlier in the year. The complaint of August 1969, initiating the suit to prevent the merger of International Telephone & Telegraph Company and Hartford Fire Insurance Co., restates those very warnings. The merger would violate the Clayton Act and lessen competition, the complaint charged, because ITT and Hartford could reciprocally trade with each other and thus "narrow" the markets of their competitors. It foresaw a defensive chain reaction of mergers:

(a) The power of ITT and Hartford to employ reciprocity in the sale of insurance will be substantially increased and the markets for Hartford's competitors will be correspondingly narrowed;

(b) Actual and future competitors of Hartford may be shut out from selling insurance to ITT and its syndicate;

(c) Actual and future independent competition by ITT and Hartford may be eliminated or diminished in numerous markets;

(d) The competitive advantages that will accrue to the defendants, which are leading firms in several industries, will raise barriers to entry and discourage smaller firms from competing in these industries;

(e) This merger will tend to trigger other mergers by companies seeking to protect themselves from the impact of this acquisition or to obtain similar competitive advantages.[21]

ITT could not have designed its acquisition of Hartford to fit more squarely into the anti-merger guidelines the attorney general had expounded to the Georgia Bar Association two months earlier. First, the combination was between two of the two hundred largest U. S. corporations. ITT, with assets of over $4 billion, was the eleventh largest industrial concern in the United States; Hartford, with assets of $1.89 billion, ranked as the 170th largest corporation.

Furthermore, the acquisition was by one of the "top two hundred manufacturing firms of a leading producer in a concentrated-industry." Hartford Fire Insurance Co. ranked "sixth among the Nation's property and liability insurance companies."

Assistant Attorney General McLaren told the assembled House Judiciary Committee on May 13, 1970, that those guidelines emphatically set forth antitrust law and the courts would not permit an ITT-Hartford merger to violate them.

> I think the present act is a very flexible instrument, and I think that were there discernible, actual or potential anticompetitive effects resulting from two of the top 200 merging that we could proceed against that. And I would be very confident that we would prevail.[22]

Dancing Partners Picked

The Justice Department filed four other suits through which it expected to restrain the current wave of mergers by securing court rulings that antitrust laws apply not only to mergers between competitors (or between customers and suppliers) but also to anticompetitive mergers of corporations not engaged in the same industry. Those suits challenged the acquisitions by:

- Ling-Temco-Vought, Inc., of Jones and Laughlin Steel Corp. (complaint filed in April 1969);
- ITT of Canteen Corp., a food supplier and operator of vending machines (complaint filed in April 1969);
- Northwest Industries of B. F. Goodrich Co. (complaint filed in May 1969); and
- ITT of Grinnell Corp., a manufacturer of fire protection equipment (complaint filed in August 1969).

The Justice Department had prepared a sixth suit to challenge the combination of Chubb Corporation and First National City Bank of New York—but won complete victory before it could even file the papers. The would-be defendants abandoned the merger upon learning of the impending litigation. The anticompetitive effects from such a merger, had it succeeded, Mr. McLaren stated, would have caused other banks to join in similar combinations to gain the same advantage. To have allowed one combination would have triggered a chain of mergers:

> We were told after we announced to counsel that we were going to proceed against the Chubb-National City Bank merger, they dropped it. Many of the banks and insurance companies had dancing partners picked out. Had we not proceeded against that particular merger, there undoubtedly would have been a regular group of mergers of this type.
>
> So I think this is something that the legislation contemplates, that the Supreme Court has talked about. It is in the legislative history. If we are intervening in a merger trend that is going to trigger further mergers that is ultimately going to result in further economic concentration, I think the present law reaches that.[23]

So confident was the assistant attorney general that his challenges would prevail in the courts, he advised Congress not to bother with enacting new legislation specifically to restrain an acquisitor that did not threaten to form a monopoly (by acquiring only in a certain industry). So strong was the government's case in those five suits, Mr. McLaren informed the House Judiciary Committee, that judicial precedent resulting from them alone should prove that the

Clayton Act without new amendment by Congress was adequate to stem the current trend of industrial concentration. He stated:

> I am inclined to defer recommending any new legislation at the present time. We will of course continue our own enforcement efforts, and I am fully confident that we will be sustained by the courts.

Judicial interpretation of the Clayton Act, he continued, already supported the Justice Department's challenge of the mergers for their effect on market structure:

> Our complaints, in every one of these cases, were based not only upon the clear legislative history of the Celler-Kefauver Amendment [to the Clayton Act] but upon well-established rules which had been announced by the Supreme Court in a series of cases in the last few years. Thus, our complaints in these five cases dealt with such matters as the elimination of potential competition; the creation of power to engage in systematic reciprocity on a large scale; the entrenchment of leading firms in concentrated markets; and the contribution to, and proliferation of, a merger trend.[24]

Emanuel Celler, House Judiciary Committee chairman, asked whether, nevertheless, antitrust laws should be strengthened:

> Very likely the success of the Celler-Kefauver Act, which prevented horizontal and vertical mergers because of reduction of competition, has probably encouraged conglomerates, where there isn't apparent competition between the

acquirer and the acquired corporation. Do you think that some legislation...is still needed so that if a...conglomerate acquired a corporation that sells or manufactures an unrelated product, unrelated to the product made by the parent company...that there must be some sort of restraint imposed by law? Or do you feel the Celler-Kefauver Act is sufficient?

Mr. McLAREN. I still think the Act is sufficient. I think that the present law is adequate to reach the major kinds of conglomerate mergers that we have proceeded against.[25]

Though of opposing political parties, the congressman and the antitrust attorney general were comrades in arms. Chairman Celler congratulated Mr. McLaren for his vigorous assertion of Congress' design. An assumption that all forces were united behind the Antitrust Division chief's resolute purpose of loosening concentration's grip, however, could be bound for disappointment. That antitrust schizophrenia was alive and well—and that Mr. McLaren's course might not be free and open—was the message of the Task Force on Productivity and Competition.

Consumers' Character Flaws

Prepared in late 1968, the document known as the Stigler Report, advised the president of the United States to follow that same course recommended by Assistant Attorney General McLaren, except in reverse, not forward: "Substantial retrenchment...in the antitrust field is highly desirable," it observed.[26] The report considered the Celler-Kefauver Amendment—which, as seen, proscribes mergers of enterprises engaged in the same industry so as to prevent monopolies—a sad mistake. The Federal Trade Commission study of 1948, which was the basis of the legislation and which indicated that the rate of industrial concentration was spiraling at a record rate, was "incompetent."[27]

Strict enforcement of consumer-protection laws, even of the Pure Food and Drug Act, won the Stigler Task Force's scorn. The act induces character flaws: "Overzealous enforcement of consumer-protection legislation can also have errant results. Elaborate requirements relating to packaging, safety, etc. can reduce the consumer's incentive to exercise care and—what is more serious—impose substantial costs on society."[28]

The Attorney General's Merger Guidelines, warning that the Justice Department would challenge the merger of any of the two hundred largest industrial corporations with any other leading enterprise in whatever industry, should have appalled the Task Force. Much more lax earlier guidelines it described as embarrassing, extraordinarily stringent, indefensible:

> The Department of Justice Merger Guidelines are extraordinarily stringent and in some aspects indefensible...[They are] so loose and unprofessional as to be positively embarrassing.[29]

The Stigler Report found that consolidations of customer with supplier, i.e., vertical combinations within the same industry (such as, hypothetically, between General Motors and DuPont), which the legislation of 1950 seeks to prevent, pose no threat to competition and even help to combat monopolies:

> On the contrary, this form of [vertical] integration whether by merger or various forms of contractual arrangements, can and does enable the integrating firm to bypass or erode monopoly elsewhere, and equally important in achieving

antitrust goals, to attain efficiencies in production and distribution.[30]

A President's Plea

To no one's surprise, the report found nothing wrong with reciprocal trading by companies operating in unrelated industries. A single sentence dismissed any concern:

> If this practice [reciprocity] leads to efficiency there is no reason to stop it; if it leads to inefficiency there is no reason why the conglomerate should adopt it (since it would reduce its profits).[31]

The nation's greatest wave of industrial concentration the Stigler Task Force took in its stride, cautioning against finding a "threat of sheer bigness to political or economic life."[32]

Rather than by enforcement of antitrust legislation, the Stigler Report recommended that competition be inspired by the words of the president of the United States urging the nation to be more competitive. It called for a nationwide address:

> We believe therefore, that the President should issue a general policy statement on competition and public regulation, to achieve at least three important purposes: To revive public support for the policy of competition...a major presidential address would be an appropriate vehicle for this declaration.[33]

A plea from the president would inspire corporations to strive harder to produce more competitive products while they fell into acquisitors' clutches. Industrial coalescence was soaring to emergency levels. "No time to wait for enactment of new laws," the Justice Department said. The government would proceed to halt the merger spiral by implementing long-standing legislation that it deemed to be fully adequate, the antitrust chief assured the committee on May 13, 1970.

Committee members and witnesses filed from the conference room that afternoon cheered by the glowing promise. The once irrepressible forces the Stigler Report embodied apparently lay at their feet. The staunchest adversaries of the department's aggressive antitrust action conceded its success.

That the Stigler Report mentality would prevail as never before was the greatest possible absurdity that day. The antitrust chiefs hardly envisioned that the Justice Department not a year later should conclude: Gigantic size exempts a corporate acquisitor from the laws of the United States (see chapter 14).

They didn't envision that acquisitors would grow so large from that exemption that the United States would despair of ever controlling them—that the United States would take counsel that it should simply identify those acquisitors too titanic to let sink and "own up to the fact that some institutions it can never let fail" regardless how inept their management (*The Financial Times*, September 13, 2009).

PART II

HOW TO BUILD AN ACQUISITOR

Acquisitors reputedly pay debt securities, such as bonds and debentures, for the companies they purchase. It's called "funny money" because it doesn't give ownership in the acquiring company. (In the 1970s it was called "Chinese money," but no more.) The stockholders of the company to be acquired, the "target," surrender their outright ownership, in the form of common stock, often in return for mere IOUs from the acquirer. In fact, however, acquisitors may pay generously in hard cash or equity securities for target company shares.

Return Favors

Four acquisitions illustrate not only such equitable methods of purchase but also methods that leave intact or even fill rather than empty the acquisitor's coffers. The first method is for the acquisitor to convince bankers that its plan to buy other corporations can benefit the bankers. Show the bank what you can do for it and you will be surprised at what the bank can do for you. The acquisitor will transfer to the bank all the banking business of the corporations it will purchase with money the bank loans. Further, the acquisition plan will enable the bank to secure millions of dollars of quick profits from various forms of stock market manipulation.

The underwriting of corporate takeovers will be so profitable to the bank that it will conceive the idea of starting another acquisition program with another diversifying company. The bank will, for example, exchange giant blocks of common stock of an airline, held in the bank's trust department, for debt securities of a gambling casino to enable the casino to take over the airline.[1] The acquisitor should not feel that it must choose a bank that appears to need more business. Rather, it should choose from the biggest. It should choose a bank run by the Rockefellers.

Shareholders Can Eat Cake

Another method is for the acquisitor to pay the owners of the target company in cash and then transfer the debt from that cash outlay to the same company upon taking control. To take control, the acquirer need not go to the expense of buying all the stock; it needs only half the stock. The owners of the other half may be startled by the sudden multimillion-dollar indebtedness. They may say that the debt has not increased the captive company's assets one cent but will diminish its earnings for the next quarter-century, if it survives. The acquisitor need only ask them why they did not sell out in time as did the other stockholders. Better, it will tell them in the annual report that the indebtedness is part of "new, imaginative debt structure."

A third method is to secure, perhaps with debt securities, a company rich in liquid reserves and use those reserves to purchase a second company. That way, without giving up any of its own capital, the acquisitor can make a $400 million purchase and afterward have $173 million more in liquid assets than it had before. It can then use those assets to purchase a second corporation from stockholders who insist on cash. In the end, the acquisitor will have spent none of its own capital and will have two new corporations and millions more in the bank than before.

The quickest acquisition method of all is to accept a call from the Pentagon some day, informing the acquisitor that the army, by granting contract price increases, has set a target company in the acquisitor's lap.

CHAPTER 4

Make Friends with a Bank

Of those four methods, Gulf & Western Industries chose to convince bankers they could profit from helping acquisitors grow. On September 10, 1965, Chase Manhattan Bank of New York loaned Gulf & Western $84 million without security, solely for the purpose of acquiring New Jersey Zinc Co. It was Gulf & Western's first major acquisition outside the automobile-parts field. New Jersey Zinc's assets of $135 million exceeded Gulf & Western's assets by over $30 million. The profits of the acquired company more than doubled the profits of the acquirer.

Charles G. Bluhdorn, the founder and CEO of Gulf & Western, came to the United States in 1943 at the age of 16, a refugee from Nazi-occupied Austria. By 1956 he had amassed a fortune from speculation over the price of coffee.

> Charles Bluhdorn, brash, excitable, and full of fire, is as he has always been, the dominant figure. The memories of his early hair-raising capers in the commodity market, which made him a millionaire in his mid-twenties, still live with him…An outsider confronted with Bluhdorn for the first time has no difficulty imagining him as the *enfant terrible* of the commodity exchange…His manner changes swiftly from persuasive explanation to table-thumping assertion, all enunciated in mile-a-minute Viennese-American…Says

Lindsay Johnson, president of New Jersey Zinc, which Gulf & Western acquired in 1966: "The more you are around Bluhdorn, the closer the moon is."[1]

In 1957 Bluhdorn had purchased the Michigan Plating & Stamping Company. The small auto-parts manufacturer was the nucleus from which Gulf & Western Industries evolved. Exactly according to the original plan, the parts manufacturer served as the basis for acquiring other corporations.

At that time the Grand Rapids firm…had two hundred employees, one aging plant, $6.5 million in annual sales, a listing on the American Stock Exchange—and, it turned out, more problems than its new young managers had realized. But as Charles G. Bluhdorn, chairman of the board of Gulf & Western, later recalled, "We couldn't sell that little bumper company, not even for $500,000 payable over five years, so we decided we had better get down to work."[2]

Fortune soon smiled. "Two shrewd young financial officers have more recently signed on at Gulf & Western: Executive Vice President Don F. Gaston and Roy T. Abbott, Jr., thirty-seven, who was a Vice President of Chase Manhattan Bank," *Fortune*, March 1968 (p. 125), reported. A few months after Chase Manhattan Bank granted Gulf & Western the unsecured $84 million loan essential to its acquisition process, Roy Abbott, the bank official who negotiated the transaction, became Gulf & Western's senior vice president. The indeed shrewd former Chase Manhattan official could then oversee Gulf & Western's performance of its return favors to the bank for making takeovers possible. He could also oversee securing from the bank more favors of the original kind. Gulf & Western's return favors, as Chase Manhattan admitted, were:

1. To enable the bank to acquire the business of competitor banks that do not use depositors' funds similarly for building acquisitors and accelerating the trend of concentration of American industry.

2. To enable the bank to profit from stock market manipulation. Gulf & Western secretly informed it of which stocks were certain to rise in value as the result of Gulf & Western's takeover attempts. The bank itself financed those attempts, both real and feigned, resulting in artificial rises in the stock prices of the target companies (see p. 4).

Return Secrets

The National Bank Act prohibits banks from acquiring control of other corporations.[3] The reason is to prevent those very favors that Chase Manhattan secured from its partnership with Gulf & Western. First, a bank could acquire its competitors' business from corporate customers sheerly by gaining control of those corporations with its depositors' money. Second, a bank could cause predictable fluctuations in market prices of companies it acquires or merely attempts to acquire.

Internal correspondence of Chase Manhattan and Gulf & Western (mainly involving the bank loan officer turned acquisitor vice president) shows that the two entities virtually formed a trust for the purpose, contrary to the law's intent, of securing unfair competitive and investment advantages through outright control of industrial enterprises. Through cooperation with Gulf & Western, Chase Manhattan Bank built an acquisitor as it could not have by itself.

1

Practically every trust created has destroyed the financial independence of some communities and of many properties; for it has centered the financing of a large part of whole lines of business in New York, and this usually with one of a few banking houses.[4]

Louis Brandeis, 1913

Local banks must not be used.[5]

Gulf & Western Policy Manual, 1967

In February 1967 officers of Chase Manhattan Bank wrote of a luncheon conversation with Gulf & Western's CEO. He promised, they wrote, that the bank would receive its due reward over its competitors throughout the country for having made the New Jersey Zinc takeover possible:

Bluhdorn said that he and the Company had not forgotten that Chase's imaginative lending had put them into business with New Jersey Zinc. He said that some day Gulf & Western will be a billion dollar company, and that Chase will figure prominently in their banking position.[6]

"This is a red letter day," bank officer Roy Abbott wrote on negotiating an early acquisition loan.[7] He advised other officers on January 28, 1964: "We are right on the threshold of a big breakthrough…Keep an eagle eye peeled for Mr. Bluhdorn." The loan officer continued: "I hope you will be on the lookout for him and roll out the red carpet when he arrives."[8]

From an appreciative bank president went a letter to an appreciative customer:[9]

February 3, 1964

DEAR MR. BLUHDORN: Tom Smith and Roy Abbott have told me the good news that Gulf & Western Industries became a customer of ours last Friday. This is wonderful news indeed because we have long hoped that we would have this additional relationship with John Duncan and yourself.

I understand that you will be coming down for lunch soon. I hope I have the pleasure of seeing you then.

Sincerely,
DAVID ROCKEFELLER

"Charlie, I'm sure we can make this very attractive to you," Abbott wrote to Bluhdorn, referring to further takeovers. "I would appreciate it if you could keep our friends in Newark out of the oil for the time being."[10] That meant to hold off taking out loans with Chase Manhattan's competitor, Prudential Life Insurance Company.

The managers of Gulf & Western's acquired subsidiaries were not always cooperative. They had to be prodded to transfer their accounts to Chase Manhattan, as the bargain for the $84 million uncollateralized loan stipulated. Chase Manhattan officers complained of "malingerers"—their name for uncooperative subsidiaries. Gulf & Western drew up instructions to its managers: "You are requested to see that the system is perpetuated each month so that our good friends at the Chase will not be constantly calling some malinger to our attention."[11] The directive to distant subsidiaries was explicit. "It is in our mutual interest for you to deposit your monthly withholding taxes with the Chase Manhattan Bank…rather than using local banks. It is considered as part of our required compensating balance."[12]

Big Rewards from Big Loans

The Bank of the Southwest, in Houston, "put up quite a howl" when Gulf & Western transferred its accounts to Chase Manhattan. But since the Texas bank happened not to grant unsecured loans to finance takeovers, it had little bargaining power for retaining that business. As a Chase Manhattan officer wrote:

> Roy Abbott [while Chase Manhattan loan officer] was told today by Gulf & Western that everything still pointed to Chase Manhattan as the prospective trustee for Gulf & Western's pension arrangements. But the Bank of the Southwest has put up quite a howl about the company's decision to give this business to us. The company has told the Bank of the Southwest that Chase Manhattan would be well-known to all of their scattered subsidiaries as a leader in this field whereas the Bank of the Southwest would be completely unknown to them.[13]

But a bank's fame, great as it might be, was not the only factor. Abbott was attentive to Chase Manhattan's interests even after his departure to Gulf & Western headquarters. Shortly before the acquisitor's purchase of South Puerto Rico Sugar Co. in March 1967, a Chase Manhattan official wrote:

In response to the branch's interest, I contacted Roy T. Abbott, Jr., of the subject's proposed parent, Gulf & Western. Roy stated that as soon as the merger is finalized he will have the accounts moved to our Puerto Rico branch. In the meantime, he does not want to make any waves until they [Puerto Rico Sugar] are in the fold [acquired].[14]

Chase Manhattan was only acquiring more business as Gulf & Western was acquiring more companies—to acquire business by fomenting the process of industrial concentration is only natural, the bank explained. To rely "continually" on loans to generate more business "is true of any banking relation," the bankers told the House Antitrust Subcommittee in August 1969:

Mr. HARKINS [Subcommittee counsel]. Did not Chase expect that its loan to Gulf & Western for the purchase of stock in New Jersey Zinc would result in a substantial increase in Chase's business?

Mr. YOUNG [Chase Manhattan Bank vice president]. I think this is true of any banking relation. Every relationship we have we continually try to develop. When we make loans, this is improving our relationship and often with the improvement of relationship in the loan area there are also improvements in other parts of the relationship, too.[15]

Unanswered is whether Chase Manhattan turns down applications for loans unlikely to generate new business. Also unanswered is how a bank retains business, if it declines to make loans for industrial consolidation, against competitors who do.

2

We are proud of our record as competitors in the American economic system. We believe that vigorous competition is essential for the preservation of that system. And the promotion of competition is what Gulf & Western Industries is about.

> David N. Judelson, President,
> Gulf & Western Industries, Inc.[16]

To acquire the business of Chase Manhattan's competitors at a faster rate, Roy Abbott (while at the bank) recommended acquisitions to Bluhdorn from confidential information obtained by the bank in its fiduciary capacities. An acquisitor's power to force suppliers of a target company to purchase from the acquisitor's subsidiaries was a major recommendation calculation. The scheme, as seen, is called "reciprocity."

By that system, suppliers that sell goods to the target company are forced to reciprocate, i.e., to purchase their goods from the acquisitor, or else lose their formerly independent customer. Because reciprocal sales are not wrought by competitive quality, the scheme subverts competitive market structure.

No Exceptions Tolerated

Reciprocity figures inspired takeover of Chicago Railroad Equipment Co. With the fiduciary letters in hand, Abbott wrote Bluhdorn: "Having in mind your reciprocity with the automobile industry and, potentially, the railroads and also some overlap into the trucking industry, I thought you might be interested in seeing some information on the Chicago Railroad Equipment Division of

AMK."[17] Subsequent recommendations also hinged on reciprocity. He did not mean that word in its literal sense, Abbott explained later: "It was a casual letter."[18]

Is Gulf & Western's policy statement against reciprocity also casual? By no means, we are told. It provides:

> Reciprocity is that practice whereby one company seeks to obtain sales of its products by agreeing to buy from a supplier provided that the supplier in turn agrees to buy from it. Reciprocity embraces every form of agreement or arrangement to that end, whether expressed or implied, direct or indirect.

> It is the policy of this Company that our purchases shall not be used as a means of effecting sales of our products. All products should be sold upon their own merits, using as inducement to the customer our own superiority in such items as price, quality, delivery, service, design, etc. In the same way, our purchases should be made with the same factors in mind, so that our needs are serviced by suppliers on the competitive merits of their own products and services.[19]

No exceptions to that policy are tolerated, David N. Judelson, president of Gulf & Western, said:

> In Gulf & Western we have a very firm policy against reciprocity. It is in our policy manual. It says the company will not indulge in reciprocity of any kind. We are very familiar with the fact that reciprocity is against the law, and we have never ever, at any time in our history, nor would we ever,

have made any acquisition from the standpoint of building up economic power in order to practice reciprocity.[20]

Freedom from Free Selection

But in utter disregard of the policy, Gulf & Western does plot acquisitions in order to secure sales independently of its products' "competitive merits." According to plan, the thousands of Gulf & Western employees would purchase insurance from to-be-acquired Associates Investment Co., not because their independent judgment of a competitive service so dictated, but because their employer so willed. E. W. Bliss. Co. would deliver its "finance business" to Associates, not because of impartial choice of a superior service but because both concerns would be members of the same syndicate. Thus assured of "tremendous potential" for noncompetitive sales, both the acquired and the acquirer would gain an advantage over competitors that had yet to join a syndicate:

Why Is Gulf & Western Interested in Associates?

...Gulf & Western has a net worth of $420,000,000 with sales of $1,200,000,000. They have about 100,000 employees. And they have other transactions pending which will increase these figures. I am sure you immediately begin to see the tremendous potential for the sale of casualty and life insurance.

In the finance area, let's look at E. W. Bliss. They make and sell a lot of machines—and we do a lot of financing of machines in our Commercial Finance area. Luckily, the machines made by Bliss are the kind we can finance; so we have access to this finance business.

Gulf & Western has made an offer for a portion of the stock of Allis-Chalmers. It is one of the largest manufacturers

of farm implements, farm equipment and other types of
heavy duty equipment. As you know, we presently do quite
a volume of business in farm equipment financing.[21]

Associates Investment Co., "the nice people with money to lend
you,"[22] with assets of $213.3 million, was Gulf & Western's largest
acquisition to date.

Illegal "block programming" similarly would free Paramount
Pictures from having to rely for sales on its consumers' free selection. By such a plan, which the Supreme Court declared violates the
Sherman Antitrust Act,[23] Paramount Pictures would sell a distributor a good film only if he also agreed to pay for a bad one. According
to a Chase Manhattan memorandum of early 1967 which records a
conversation between Roy Abbott (while Gulf & Western vice president) and the vice president of Paramount Pictures, the distributor
would have to pay for "B" pictures if he wanted any "A" pictures:

> While block programming is not legally allowed (block
> programming is where the movie studio will make the film
> exhibitor take some "B" pictures otherwise he can't get the
> "A" pictures) it is nevertheless a factor in getting a com-
> pany's pictures shown.[24]

Gulf & Western Chief Executive Bluhdorn devised the contract
for delivering to television viewers the otherwise unsalable "freight
cars." He described the plan while at lunch in February 1967 with
Roy Abbott and Chase Manhattan officials, whose notes record:

> Bluhdorn discussed the purchase of Paramount which he
> felt was equivalent to buying a "bank." In his most recent
> lease package negotiations, he worked out a deal where

the network had to take "locomotives" (that commanded record box office performances), "freight cars" (so-called "B" films) and reruns. As he puts it, this market of network leasing is definitely a seller's market.[25]

With that intelligence, Chase Manhattan officials at the luncheon table confidently contributed $10 million of the $30 million loan for Gulf & Western's purchase of Paramount Pictures.

Customer Satisfaction Paramount

On August 7, 1969, at the moment the "block programming" memorandums were being placed in the record of the House Antitrust Subcommittee investigation, former White House assistant and counsel for Gulf & Western, Joseph A. Califano, entered the hearing room to deny that the acquisitor had put the "freight car" scheme into practice.[26] As evidence he asked that the subcommittee accept for the record a copy of a contract clause and statements of policy against "block programming"—policy similar to Gulf & Western's formal expression against reciprocity. He never denied, however, that the anticompetitive programming scheme inspired, as the memorandums indicate, the Paramount Pictures acquisition.

"Block programming" would deny customers free choice of television programs. But Gulf & Western announced: "G&W's emphasis on customer satisfaction is in the oldest tradition of American business."[27]

3

Public knowledge of a proposed acquisition usually causes the market price of the target company to rise. The acquirer must pay more than the market price to induce the owners to sell. Investors thus raise their bid for the target company stock to the limit they believe the acquirer will pay. Investors with "inside" or confidential information of a proposed merger, therefore, may purchase shares of the target company with the accurate expectation of a quick rise in market value as soon as the acquirer makes a public announcement of the merger plan.

The House Antitrust Subcommittee counsel asked Mr. Bluhdorn whether, as to be expected, stocks of target companies rose after Gulf & Western merger agreements became known to the public:

> Mr. HARKINS. In the course of your acquisition program, Mr. Bluhdorn, have you observed that when the public became aware that Gulf & Western was considering an acquisition of a company, the stock in that company rose?

> Mr. BLUHDORN. I would say that whenever possible, we went to great lengths to keep these things of such confidential nature for that very reason. However, sometimes in the situations that we are dealing with, we must go to our attorneys continuously, as we had to do, for instance, with Simpson, Thacher. At times we were obligated to make announcements, but we only did so when [we] were required to under legal procedures.

> Mr. HARKINS. But my question was: When the announcement was made, or when it became public information that

you were considering acquiring a particular company, did the price of the stock in that company rise?

Mr. BLUHDORN. I couldn't say. For instance, in Paramount, I would recollect when it was reported the stock went down.

THE CHAIRMAN [Emanuel Celler of New York]. It is a general question, Mr. Bluhdorn.

Mr. BLUHDORN. Sometimes, yes.

The CHAIRMAN. You don't have to give an explanation. Did the stock of the acquired companies usually rise?

Mr. BLUHDORN. More often than not, Mr. Chairman. I cannot give a "yes" or "no" answer. More often than not.[28]

In fact, the stock prices of all thirteen major Gulf & Western acquisitions rose sharply during the period of confidential merger negotiations, as Table 1 shows. The stock prices of eight of those companies rose over 25% during the period of a month (at most) before public announcement of merger. With that period extended to the date of consummation of the merger, the price of eight of the thirteen largest target companies rose over 50%.

Table 1

Stock Prices of Major Takeovers Financed by Chase Manhattan Bank, Acquirer of J.P. Morgan Co. in 2000—Rescued with $43 Billion of Public Funds in 2009

	Price range during period from date of press release to same date of previous month (1)	*Price on date of 1st press release (2)*	*Price on date of merger (3)*	*Percent change from the lower price in col. 1 to col. 2 (4)*	*Percent change from col. 2 to col. 3 (5)*	*Percent change from the lower price in col. 1 to col. 3 (6)*
Crampton Manufacturing Co.	$3 7/8 to $4	$ 4 7/8	$ 6	25.8	23.1	54.8
Miller Manufacturing Co.	$10 to $10 5/8	13	14 3/8	30.0	10.5	43.7
New Jersey Zinc Co.	$32 3/4 to $38 1/2	38 1/2	51 1/4	17.6	33.1	56.5
Paramount Pictures Corp.	$65 5/8 to $81 3/4	81	75 1/4	23.4	(7.1)	14.7
North & Judd Manufacturing Co.	$33 to $35	35	38	6.1	8.6	15.1
Collyer Insulated Wire Co.	$32 1/2 to $44 1/2	43	53	32.3	23.3	63.3
Desilu Productions, Inc.	$9 to $11 5/8	15	14 5/8	66.7	(2.5)	62.5
South Puerto Rico Sugar Co.	$23 1/2 to $33	33	50	40.4	51.5	127.6
E. W. Bliss Co.	$23 3/4 to $28 1/4	27 1/4	34 1/8	14.7	25.2	43.7
Consolidated Cigar Corp.	$20 5/8 to $25 3/8	23 5/8	32	14.5	35.4	55.2
Universal American Corp.	$18 to $24	24	31 1/8	33.3	29.7	72.9
Brown Co.	$18 7/8 to $25 7/8	25 7/8	19 5/8	37.1	(24.2)	4.0
Associates Investment Co.	$25 1/4 to $33 3/8	33 3/8	38 1/2	32.2	15.4	52.5

SOURCE: *Hearings on Conglomerate Corporations*, Part 1, p.105.

Table 2

Common Stock Prices of Gulf & Western's Major Abandoned Transactions[1]

	Price range during period from date of press release to same date of previous month (1)	*Price on date of 1st press release (2)*	*Price at which G.&W. sold upon abandonment of acquisition attempt (3)*	*Percent change from – Lower price in col. 1 to col. 2 (4)*	*Col. 2 to col. 3[2] (5)*	*Lower price in col. 1 to col. 3 (6)*	*Announcement date[3] (7)*	*Abandonment date[3] (8)*
Universal American Corp.	$11 3/4–$17 1/2	$17 3/4	$ 12 3/4	47.9	(26.6)	8.5	Apr. 14, 1966	June 1, 1966
Armour & Co.	34 1/2– 44 1/2	44 1/2	36 3/8	29.0	(18.3)	5.4	Jan. 16, 1968	Feb. 5, 1968
Allis-Chalmers Manufacturing Co.	29 5/8– 33 5/8	33 1/4	32	13.2	(3.8)	8.0	May 7, 1968	Dec. 6, 1968
Pan American World Airways	20 5/8– 23 7/8	23 7/8	29 7/8	15.8	25.1	44.8	Sept. 16, 1968	Jan. 9, 1969
Sinclair Oil Corp.	76 1/2– 90 5/8	90 5/8	130	18.5	43.4	69.9	Oct. 24, 1968	Dec. 31, 1968

1. From the Investment Statistics Laboratory Daily Price Index for the New York Stock Exchange
2. Parentheses indicate percentage decrease.
3. From Gulf & Western documents.

SOURCE: *Conglomerate Investigation Report*, p.197.

Column 1 in Table 2 shows the lowest price investors with "inside" information of pending merger could have paid for the stock of Gulf & Western target companies during the month before that information was made public. Column 3 shows the height to which the stock had risen by the date of merger (after public announcement). Column 6 shows the percentage of the appreciation, i.e., the profit to be gained from having purchased the stock a month (at most) before public disclosure of each merger plan.

The sudden rise in market values of the stock of all thirteen largest target companies before the merger plans became public knowledge strongly indicates that investors with that knowledge were purchasing the stock (in anticipation of its increased value) and thus raising the market prices. The holders from whom the investors purchased before the price rose of course would not have sold had they likewise possessed confidential information of the pending acquisition.

Gulf & Western disagreed, contending that Table 2 cannot indicate that certain investors purchased stock on the basis of confidential information from the acquisitor, and thereby deprived the former holders of the foreseeable gain, because they "abound" with errors and because "stock prices constantly fluctuate":

Gulf & Western believes that it is impossible to draw any valid conclusions on the basis of [Tables 1 and 2] because:

1. No consideration is given to the fact that stock prices constantly fluctuate as a result of a myriad of factors, only one of which is public announcement of a possible merger;
2. No consideration is given to the special circumstances surrounding certain of the enumerated transactions; and
3. Clerical errors with resulting inaccurate computations abound in the charts.[29]

The prices shown in Tables 1 and 2, however, can hardly be described as fluctuating. They consistently rise. Also, after submitting suggestions for minor changes reflected in the tables as here shown, Gulf & Western withdrew its charge of inaccuracy.[30] An investor that regularly received confidential disclosures of Gulf & Western's proposed target companies was Chase Manhattan Bank—the very financier of the acquisition program that would cause the predictable rise in the price of those merger targets.

Better Inside Information

A Chase Manhattan memorandum of May 9, 1966, describes the duty of the former bank loan officer, Roy Abbott, now Bluhdorn's vice president, to inform the bank of which companies Gulf & Western proposed to acquire before the public knew. "By notifying us [Chase Manhattan] prior to [merger] announcement in the newspapers," Abbott would enable his former employer bank to purchase stock of the target companies before it rose upon public announcement of merger.

From Chase Manhattan Bank files came the memo:

MAY 9, 1966.

CONFIDENTIAL
CREDIT FILES, GULF & WESTERN INDUSTRIES, INC.,
HOUSTON, TEX.
NEW JERSEY ZINC CO., NEW YORK, N.Y.

Roy Abbott stopped in late this afternoon to discuss various aspects of both Gulf & Western and New Jersey Zinc, with Tom Hill, Harold Young and myself... Incidentally, Roy also agreed to keep us better informed concerning proposed mergers and acquisitions and important investments, i.e. notifying us prior to announcements in the newspapers.[31]

The counsel demanded to know whether Roy Abbott did in fact, as the memorandum indicates, inform his former employer of pending Gulf & Western acquisitions before the predictable rise in value:

Mr. HARKINS. Mr. Abbott, as an official of Gulf & Western, was it your duty to keep Chase informed of proposed mergers and acquisitions?

Mr. ABBOTT. It was, to the extent that they either were going to affect the company significantly as far as its financial situation or projections were concerned, or to the extent it may have affected any loan agreements or any written legal arrangements we had with them.

Mr. HARKINS. Did you notify Chase prior to the announcement of an acquisition in the newspapers?

Mr. ABBOTT. Occasionally, yes. It would have been much the same as if I had gone to a neighbor and said, "Look, I am going to extend my house, and [it's] going to cost about $5,000, can you lend me $6,000 to be on the safe side," and three days later he would look out the window and see a new car.

Mr. HARKINS. How is it like that?

Mr. ABBOTT. If he would look out the window subsequently and see I had a new car, my neighbor would be very upset. Very much in the same light we keep our banks informed of what we are doing as it will relate to our financial condition; obviously, we have loans, we have lending arrangements, and they are conditioned on what we have said we are going to be doing.

If our plans change, obviously these are important facts which the bank has to know.

Mr. HARKINS. Have you finished your explanation, Mr. Abbott?

Mr. ABBOTT. I have.

Mr. HARKINS. Did you agree to notify Chase prior to the announcement in the newspapers?

Mr. ABBOTT. I am sorry?

Mr. HARKINS. Did you agree to notify Chase of proposed mergers of Gulf & Western, prior to their announcement in the newspapers?

Mr. ABBOTT. If they would significantly affect the areas I just had mentioned.[32]

Chase Manhattan officials responded more directly:

> Mr. HARKINS. After Mr. Abbott went to Gulf & Western, he supplied information to Chase about Gulf & Western's acquisitions; did he not?
>
> Mr. YOUNG. Yes, sir.
>
> Mr. HARKINS. Was this information he supplied given to Chase officials prior to public announcement of the mergers?
>
> Mr. YOUNG. In most cases I would say yes, sir.
>
> Mr. HARKINS. And as members of the credit department of Chase, did you expect Mr. Abbott to provide this type of information to you?
>
> Mr. YOUNG. Yes, sir; we did. [33]

Gulf & Western, according to those admissions, had passed to Chase Manhattan confidential information of which stocks its own acquisition loans would cause to rise. The bank was enabled, then, to manipulate the stock market in its favor through its power to lend other people's money.

In Full Appreciation

Further internal correspondence attests to Chase Manhattan's persistence in regularly securing Gulf & Western's confidential plans of merger. A memorandum from a Chase Manhattan vice president of April 22, 1966, recorded a recent disclosure:

Roy Abbott called to let us know that another proposed acquisition will be announced probably on Monday, April 25th. The company's name is Muntz Stereo Pak, Inc., which is in the cartridge tape business. This is the same company our Credit Department checked on recently for G. & W. The acquisition, if consummated, will be for stock.[34]

Gulf & Western Vice President Roy Abbott, in full appreciation of a loan of $15 million, would make sure that the Gulf & Western financial officer "discreetly" apprised the bank of the acquisitor's "acquisition plans," the writer concluded. Abbott wanted "us to let him know if we" needed more information:

Also had a general conversation on the advisability of Gulf & Western keeping all of its banks up to date on acquisition plans. Roy, of course, fully appreciates the problems from our side of the desk at the bank and said this is certainly their objective. However, he feels this function is one which belongs to Herb Neyland and will discreetly cover the area with him at an early opportunity. Roy also fully appreciates the position we have taken on the proposed $15,000,000 line of credit and wants us to let him know if we do not get everything we want from Herb.[35]

The same Gulf & Western financial official, a Chase Manhattan memorandum of February 23, 1967, recorded, "confidentially" apprised the bank of subsequent acquisition plans:

6. Acquisitions…

He said that they would probably back out of the E. W. Bliss situation. Bliss' management has not been receptive at all to the possibility of a merger.

He mentioned that they had a merger and acquisition team which was constantly screening possibilities. At present he said they had three which are, confidentially, Panavision, Gulf Forge, and Gulf American Land Development Company.[36]

Profit from Credulity

Actual acquisition is by no means essential to causing a rise in the stock price of a reputed target company. All that is necessary is to create the appearance of a takeover attempt by purchasing large blocks of a company's stock and issuing a press release announcing intent to acquire. Public investors expect the reputed acquirer to pay a premium over the market price for more stock. Consequently, they bid higher and higher. The reputed acquirer can then capitalize on public credulity by selling while its merger rumors persist and the price is still high. The decline comes after the spurious acquirer has safely liquidated its holdings.

From one such spurious acquisition attempt or "abandoned transaction"—the sale of stock of Armour & Co.—Gulf & Western gained $16 million in less than a year. It sold before the public learned that its managers had abandoned plans for merger. From four such transactions, it gained over $51 million (see Table 3).

Table 3

Gulf & Western Abandoned Transactions and Profits

	Cost of stock to G.&W. before it announced intent to merge	Proceeds received by G.&W. from sale of stock before announcing abandonment of merger plans and while public investors thought merger would occur	Profit	Date of transaction
Armour & Co.	$ 28,149,915	$ 44,400,000	$16,250,085	Jan. 16, 1968 to Oct. 15, 1968.
Allis-Chalmers Manufacturing Co.	117,074,000	122,080,000	5,006,000	May 7, 1968 to Dec. 6, 1968.
Sinclair Oil Corp.	88,774,283	112,953,620	24,179,337	Oct. 24, 1968 to Mar. 4, 1969.
Pan American World Airways, Inc.	10,802,844	17,250,000	6,447,156	January 1968 to Apr. 17, 1969.
Total profit			$51,882,578	

SOURCE: *Hearings on Conglomerate Corporations*, Part 1, p.110.

As it had financed purchases of real acquisition targets, Chase Manhattan financed Gulf & Western's profitable purchases of shares of the "abandoned targets."

The bank also learned from its former loan officer, before disclosure to the public, of the imminence of reports of rapidly increased Gulf & Western earnings. The reports resulted from nothing more than changes in the acquired companies' methods of accounting. New Jersey Zinc, for instance, Mr. Abbott reported to Chase Manhattan on September 28, 1966, would enjoy a $3.65 million profit increase (to be reported later) resulting entirely from a change in accounting for depreciation.[37] From that not-yet-public information of nonrecurring artificial gain any "insider" could expect an increase in market price and, hence, buy before it occurred.

Disclosure Feared

A similar accounting change, which likewise provided figures that indicated improved performance of an acquired company under Gulf & Western management, resulted in a showing of $641,000 nonrecurring increased earnings for Consolidated Cigar Corp. in

1967.[38] Still another change of depreciation accounting after merger enabled the Gulf & Western managers to report a $120,000 profit increase for Universal American Corporation.[39]

Gulf & Western's automobile-parts division, a company vice president informed Bluhdorn on June 6, 1964, would report earnings of $1.230 million. One million of that, he continued, would come not from "true earnings" but rather from Gulf & Western's changes of inventory accounting methods of the acquired Beard & Stone Electric Co. and from the sale of real estate. The real earnings of the parts companies are "virtually nil" he continued, "and I am extremely fearful of any detailed disclosures...":

> The consolidated statement of earnings for the nine months ended April 30, shows that the automotive parts subsidiaries have made $1,123,000 before taxes. Of this amount more than a million dollars represents "special items" such as inventory bargain, taking Beard & Stone off LIFO [last in, first out], gain on sale of real estate at Reading, etc.; thus the true earnings of the parts companies are virtually at nil for this year and I am extremely fearful of any detailed disclosures we might have to make in a registration statement.[40]

Chase Manhattan officials denied that the bank profited from the receipt of "inside" information by purchasing the stock that its acquisition loans later caused to rise. Just as Gulf & Western had pointed to company policy statements as proof that it refrained from reciprocal trading, so Chase Manhattan pointed to public declarations that it looked askance on any idea of using confidential "inside" information for deciding on stock purchases. The declaration of May 21, 1965, instructed all Chase Manhattan officers not to use confidential information for the bank's or for their own advantage:

> The lessons to be learned from the Texas Gulf case must not be forgotten. Inside information gained by the Bank or any employee should not be used by the Bank or any employee to obtain an investment advantage. It is equally clear that such inside information should not be disclosed to persons outside the Bank since they might make improper use of the information or might disclose it to other persons who might make improper use of the information. It follows from these principles that inside information obtained from a corporation or its directors, officers, employees or other insiders is not to be used by the Bank or employees in purchasing or selling securities, or recommending to others the purchase or sale of securities.[41]

Still another declaration, of November 4, 1968, again referring to the Texas Gulf Sulfur case, decided two months earlier, stated that the bank must not use information that it does not make public. To purchase securities on the basis of such "inside" information, the bank warned its employees, amounts to fraud:

> The Court of Appeals has indicated that the anti-fraud provisions (Section 10 (b) of the Securities Exchange Act and SEC rule 10b-5) are applicable to any person in possession of material inside information even though such person may not be an insider. The court said "... *anyone* in possession of material inside information must either disclose it or... must abstain from *trading in or recommending* the

securities concerned while such inside information remains undisclosed."[42]

A Wall Between Departments

"Inside" information must not flow between bank departments, the declaration continued:

> To insure the proper use and control of information received by the Bank in its several capacities there is to be no flow, or incidental communication, of inside information regarding other companies from the commercial departments or divisions of the Bank to the Fiduciary Investment Department or to the Pension or Personal Trust Divisions of the Trust Department. Similarly, there is to be no flow, or incidental communication, of inside information regarding other companies from the Fiduciary Investment Department or the Pension or Personal Trust Divisions of the Trust Department to other departments and divisions of the Bank.[43]

To implement that policy, "we have very definitely…established a wall between the commercial department and other departments of the Bank," Chase Manhattan's counsel assured the House Antitrust Committee.[44]

The bank, however, never showed that the figurative wall existed other than by saying that it did and by referring to the policy statements. The office memorandums written over a period of years prove that disclosure of acquisition targets was disseminated, as the bank admitted, to many of its employees. "No mechanism for control [i.e., prevention of access to the data by bank divisions which purchase securities]," the House Antitrust Subcommittee report states, "was demonstrated."[45]

Make It Clear

Chase Manhattan's regular insistence on confidential disclosures of Gulf & Western's target companies, its complaints to the acquisitor whenever disclosures were not prompt, and the careful recording of the disclosures belie its vaunted claims of self-restraint.

The bank actually instructed its officers:

> In addition, trust and fiduciary investment personnel, when interviewing corporate officers or representatives of brokerage houses, should make it clear that they are not seeking [for investment purpose] inside information.[46]

But the officers' eagerness (revealed by the documents) to learn of Gulf & Western's next acquisition moves simply does not mesh with that instruction.

Gulf & Western's annual reports to the public during the last decade show rocketing increases in resources and earnings. During the ten-year period 1960-1969 those reports indicate that total assets increased by 18,743%, total revenues by 6,608%, net income by 1,674%, common stock equity by 9,057%, and net worth by 9,847% (see *Conglomerate Investigation Report*, p. 188).

The drastic increases are the simple result of adding each year the assets and revenues of acquired companies to Gulf & Western assets and revenues. In 1968 "external" sales and profits, i.e., amounts contributed by companies acquired that year, comprise 83.4% and 81.8%, respectively, of Gulf & Western's annual increase of sales and profits. Hence, had Gulf & Western acquired no companies in 1968, i.e., had its growth been entirely "internal," the figures for growth of sales and profits would have been but 16.6% and 18.2%, respectively, of the figures reported (see *Conglomerate Investigation Report*, p. 190).

Motorized Driftwood

To show growth by purchasing revenues is costly. For every dollar of its net worth, Gulf & Western owed 30 cents in 1960, but owed $1.50 in 1969. In 1960, Gulf & Western could pay interest charges on its indebtedness seven times over from its earnings. By 1969, however, more than *half* the acquisitor's earnings went for interest payments.

Table 4

Gulf & Western Industries, Inc., and Subsidiaries, Selected Financial Stability Ratios

	1960	1965	1966	1967	1968	1969
Long-term debt/net worth[1]	0.30	0.80	0.95	1.1	1.5	1.5
Interest coverage	7.0	6.4	5.1	7.2	4.9	1.9
Interest and preferred dividend coverage	7.0	6.4	3.1	3.6	3.3	1.5

1. Includes minority interest.
SOURCE: *Conglomerate Investigation Report*, p.189; prepared by the House Antitrust Subcommittee staff from annual reports of Gulf & Western Industries, Inc., and its subsidiaries.

In *Fortune,* March 1968 (p.204), W. S. Rukeyser reasoned that "to justify itself economically, a conglomerate must demonstrate ability to improve the performance of the companies it acquires." Charles Bluhdorn agreed with him in the same article. "Our job is to motorize the driftwood," the acquisitor said. "Driftwood" was his word for the companies Gulf & Western acquires.

A comparison of pre-merger performance with post-merger performance shows whether the purchased companies were in fact driftwood before merger and whether the acquisitor motorized them. Profitability figures secured by the House Antitrust Subcommittee through the year 1969 for eleven of the thirteen major Gulf & Western acquired corporations—90% of G&W's

total acquired assets—for the first time provided data for such a scrutiny. Comparison of performance after merger with independent performance shows that eight of the eleven companies earned considerably less in proportion to their assets under Gulf & Western control than under independent management (see *Conglomerate Investigation Report,* p. 188).

Gulf & Western chose not to reveal to the subcommittee performance data for years after 1969. Congress chose not to subpoena it.

Chase Manhattan Bank skirted the rule against risking depositors' money for speculation, in return for the acquisitor's hot market tips the depositors' money provided. But after the scrapping of the ITT-Hartford suit, Congress scrapped the rule. Chase Manhattan was free to become an investor itself by taking over J.P. Morgan Co. and purchasing Madoff securities for its customers. After the public supplied $43 billion to bail it out, the combined bank helped obscure Greece's ruinous finances. Then it purchased credit-default swaps to bet on the nation's ruin (see p. 5).

CHAPTER 5

Shift the Purchase Debt to the Purchased Company

To the victor belong the spoils, and to the acquired belongs the acquisition debt. James J. Ling attended the Naval Electrical School, saved $2,000 of his service pay, and at the end of World War II set out as an electrical contractor in Dallas. He sold stock in his company from a stand one summer at the Texas State Fair. In 1961 Ling Electrical Co. acquired the aircraft manufacturer Chance Vought Corp. and changed its name to Ling-Temco-Vought, Inc. By 1964 LTV was 186th on *Fortune's* list of the largest U. S. industrial corporations. By 1969 it was the twenty-fifth largest, producing in the fields of aerospace (aircraft, missiles, space maneuvering units), meat and foods, military electronics (reconnaissance and intelligence systems), wire and cable, floor covering, commercial electronics, traffic control systems, recreation and athletics, pharmaceuticals and chemicals, passenger and freight air transportation, and steel. During the five years ending in 1969, consolidated sales increased 800 percent, and consolidated assets, almost entirely acquired, increased 2,300 percent.[1]

Industry Ignorance

In mid-1966, when LTV produced mainly in the fields of aerospace, wire and cable, and electronics, it began consideration of its greatest acquisition ambition (soon to be dwarfed): Wilson & Co., one of the nation's most profitable meat packagers. With assets of over $195 million, it surpassed in size and revenue all the companies on which LTV had previously set its sights. Wilson & Co. operated eleven meat-packing plants in Iowa, Minnesota, Kansas, Nebraska, Massachusetts, California, Oklahoma, Utah, Maryland, and Colorado. One division, Wilson Athletic Goods Mfg. Co., was the largest producer of athletic goods in the United States. In fourteen factories, the division produced golf, tennis, baseball, football, and basketball equipment. Sales branches and warehouses were located in thirty-one cities. Wilson Chemical Industries, a second division, produced pharmaceutical supplies from animal sources, mainly in Illinois and Pennsylvania. The total revenue of Wilson & Co. surpassed $990 million.

The acquisitor considered not what it could contribute to the target company, but rather what Wilson and its "excellent" management could contribute to LTV. The acquisitor had no inclination to enter a field of industry that related to LTV's capability—nor did it look for a field in which it might heighten competition. Rather, it looked for a well-functioning company that could increase LTV's profits and be had for a good price.

"We make no claim to understand the industry," LTV stated unabashedly. To "diversify with companies that have above average management" was the objective avowed to LTV directors in December 1966. "Wilson's management is doing an excellent job. We believe management is the key to any successful diversification effort and our investment in Wilson is a vote of confidence in the people who make up this company."[2] The presentation to the LTV directors did not explain a hundred-years-old enterprise's sudden need for a parent management.

Nor did it explain the logic by which the acquirer, making no claim to understand the industry much less to perform and compete in it, expected Wilson & Co. to improve under LTV control. Those questions were beside the point. In the first place, there was very little room for improvement, according to LTV's "Reasons for Interest" in Wilson & Co., also presented to the board:[3]

QUALITY COMPANY
Established 185—third largest in industry

GROWTH IN SALES AND EARNINGS
Sales: 10-year annual growth rate of 4.5 per cent
Earnings: 10-year annual growth rate of 4.0 per cent
Current Ratio 2.6
Working Capital $64.6 million
Total Debt only $27.2 million
Long term debt to equity ratio only O. 16/1

GOOD RELATION BETWEEN BOOK VALUE
AND TENDER PRICE
Book Value $44

Coupled with LTV's high regard for Wilson & Co. was its calculation—based on observation of fluctuating hog prices—that time was ripe for a short-term rise in meat-packing profits. There would be a good chance, then, for showing an earnings increase right after acquisition. Hogs were in "long supply," James Ling was informed, and probably would be for two more years. And investors would be all the more pleased when those expected profit increases were accentuated by certain inventory accounting techniques.[4]

No less a consideration than the quality of the company and the market condition of the inventory was that LTV could pay more than the market price for Wilson shares and still report increased earnings per share after combination or "pooling" of the earnings of the two companies. The price of Wilson shares was low in comparison to earnings. Consequently, those shares represented greater earnings in proportion to market price than did LTV shares. Thus, Wilson shareholders who took LTV stock in exchange (rather than cash, as did those who accepted the tender offer) received a security earning less profit than the one they had surrendered.[5]

Retirement funds of LTV employee were a source of funds for its purchase of Wilson shares. The fund bought $5 million of LTV notes (IOUs). The proceeds went to purchase the Wilson stock. The trustee, Republic National Bank of Dallas, objected to spending LTV employee funds for takeovers. Ling's lieutenants informed him of the "problem with Republic National" and that "the incidents… have had a serious effect on the operating relationships between the LTV Investment Committee and the Republic Bank as trustee." Nevertheless, the committee did approve the purchase of $5 million of those notes. In December of that year, LTV fired uncooperative Republic National Bank as retirement fund trustee.[6]

Curiously, asked during congressional hearings whether he thought LTV could properly use LTV employee trust funds for LTV acquisitions, James Ling replied, "Not whatsoever."[7]

The bulk of the acquisition funds, according to Ling's first plan, would come from Bank of America.[8] With that loan LTV would purchase over half the amount of stock needed for control of Wilson from mutual funds and other institutional investors. But Bank of America did not share Chase Manhattan Bank's enthusiasm for helping acquisitors grow. The bank would not grant even a secured loan for acquisition—if "secured" is the word for a loan (as envisioned) collateralized by the very same stock it is used to purchase. Interest rates were soaring, and the Federal Reserve Board's policy was to restrict credit.

Crisis in Reverse

Ling conceived of borrowing, at a lower rate, U. S. dollars held in Europe, or Eurodollars. It would work in reverse to the balance of payments crisis, resulting in the largest sum ever then to be repatriated. The U. S. secretary of treasury, forgetting that the dollars would have to be paid back to the Europeans, was delighted.[9]

On December 7, 1966, the president of LTV wrote to LTV director W. H. Osborn, an official of the investment banker, Lehman Brothers:

> The banks are extremely interested in our Eurodollar program and the uniqueness of this approach. They indicated a hundred percent support for it; in fact, I believe they were enthusiastic, particularly for bankers. I did not mention the name of the company involved, although there was a substantial interest in this; in fact, many implications could have been taken that they wanted to know the specific name.[10]

Rothschilds and Lehman Brothers then began selling the plan to European bankers who they hoped could locate sixty million expatriated dollars. The banks would take as security for the loan the same Wilson stock that LTV purchased with the loan proceeds. Although the stock would be out of LTV's possession, LTV would nevertheless still enjoy the voting rights. The acquisitor would gain a controlling interest in Wilson & Co. just as if it owned the Wilson stock outright.

Borrowing from Borrowings

By that logic, a stock purchaser with a sum of money infinitely smaller than a corporation's total value might acquire that corporation by successively purchasing its stock with successive loans secured by each stock purchase. By putting that theory into practice, LTV purchased Wilson & Co.

Through use of funds borrowed by LTV, in turn borrowed by the banks from the public, LTV acquired over $196 million in assets from that single acquisition without issuing one new share of common stock. LTV reported earnings per share then by dividing all its combined earnings only by the number of common—not preferred—shares. But upon assuming control of the meat packager, LTV canceled the common shares of Wilson holders who had not accepted the tender offer to sell for cash. In payment for cancellation, the acquisitor issued them LTV preferred shares. Thus, the earnings of the two companies as combined after merger were divided by a smaller divisor (number of common shares) than those earnings had been before merger. The result was an astronomical increase in the reported LTV earnings per share—and, consequently, in the market evaluation of LTV shares—and, consequently, in LTV's resources for taking over more corporations. The acquisitor's earnings rose from $13.683 million in 1966 to $34,003 million in 1967 after the merger. Sixteen million dollars of that $21 million earnings increase (resulting from other acquisitions as well) came from Wilson's addition.[11]

Dawn of New Era

LTV attributed the rising earnings to its allegedly superior management ability. The increase demonstrated, its management asserted, that through concentration of corporate assets there had dawned a new era of corporate profitability.[12]

After the report of greater earnings per share, LTV common stock rose from a high of 52 in the fourth quarter of 1966 to a high of 166 less than a year later (see *LTV Annual Reports*). A rising market price assures successful negotiation for more loans for more industrial combinations.

The only possible glitch in the cycle is any failure of the new assets under the parent company's control to generate sufficient earnings. If those assets should provide less income than necessary to stimulate a rise in the parent company's stock, the acquisition process breaks down. Worse, if the acquired company's operations do not generate proceeds adequate even to pay the astronomical acquisition debt and interest, or—more unmentionable—result in a loss, havoc ensues.

But elation prevailed in late December 1966 when word came across the Atlantic to the LTV Tower in Dallas that commitments of Eurodollars by lenders had surpassed $42 million (see *Hearings on Conglomerate Corporations*, Part 6, p. 560, prepared by the House Judiciary Committee Staff).

The moment to strike was at hand. On Tuesday morning, December 21, Wilson shareholders across the nation unfolded their newspapers and learned the splendid news. Almost overnight their investment had increased $15 a share above the market price. LTV was offering to pay, they read, $62.50 for each of their common shares selling shortly before for $47.00.

Nothing Succeeds Like Success

In turn, the European bankers soon learned that the shareholders were accepting the offer (of a little less than 33% cash profit on the basis of pre-merger market price). The banks immediately consented to increase the loan limit by $20 million, to discard various security provisions, and to send over $8 million more Eurodollars immediately.

LTV President Clyde Skeen wrote the vice president for finance, on January 11, 1967:

With respect to the final word on the consents in Europe—he [Osborne, the Lehman partner-LTV director] had just heard from Phillip Shelbourne and all banks had consented with the exception of Schroeders in London and BCI in Italy—and they anticipated these consents before the day is out, thereby making the consents unanimous. Both Bill and Shelbourne were elated by these events.[13]

Thirty million more dollars came from the "private placement" subscribed to by domestic lenders, including the LTV Employee Retirement Trust. With the $80 million borrowing, LTV had purchased 53% of Wilson voting stock, paying a premium of $19.5 million over the market price.

Purposes Unknown

The acquirer, purchasing directly from the target company shareholders, never negotiated with Wilson's management. As takeover became imminent, however, LTV Treasurer B. L. Brown courteously sent the Wilson directors some literature about the acquisitor that in a matter of days would put them out on the street. Director Thomas B. Freeman replied on Christmas Day that he had already apprised himself of LTV.[14]

DECEMBER 22, 1966.

Mr. THOMAS B. FREEMAN,
Tucson, Ariz.

DEAR MR. FREEMAN: At the request of Mr. James J. Ling, the enclosed material on Ling-Temco-Vought, Inc. is being forwarded to you. Similar material was given to some of the directors of Wilson & Co., Inc. but regrettably it was not possible to contact each of you personally in order to acquaint you with our company.

Sincerely,
B. L. BROWN.

Tuscon, Ariz., December 25, 1966.
Mr. B. L. BROWN,
Treasurer, Ling-Temco-Vought, Inc.,
Dallas, Tex.

DEAR SIR: This is to acknowledge receipt of the LTV Material you sent me. However, it was a bit late as I naturally had immediately obtained adequate information about sales, operation results and balance sheets of your company and its affiliates, after learning from the press your company's announced intentions earlier this week.

It is apparent and fully recognized by me that you and your associates hold some, if not all of Wilson and Company's Directors in contempt and that you thought them beneath your consideration in laying your plans. It is also apparent that you and they intend to obtain control of Wilson and Company, Inc., if you can and for the purposes unknown to me.

From many past experiences with operators and operations such as you and yours seem to be, my opinion is that Wilson's competent and enthusiastic management we have been developing for the last thirteen years will quickly be demoralized and will lose much of its effectiveness in any event.

The fine line between profit and loss in the meat end of the business is such that a sizable profit can easily become a serious loss. We have ample evidence of that when we look at what has just happened to the meat operations of Rath, Cudahy, Hygrade and others. If you want to go into the meat business, you can acquire any or all of these at bargain prices.

A very large percentage of Wilson's sales are to consumers in the meat category. They benefit little from long profit, war inspired sales to the Army, the Navy, the Air Force or from other governmental projects.

Yours truly,
THOMAS B. FREEMAN.

Roscoe G. Haynie, president of Wilson & Co., Inc., received word from his new overlord on January 12:

You are hereby advised that as of January 5, 1967, Ling-Temco-Vought, Inc., owned beneficially 1,295,597 shares of common stock of Wilson & Co., Inc. Suggest you amend proxy statement accordingly.

CLYDE SKEEN, President, Ling-Temco-Vought, Inc.[15]

LTV then had control of Wilson. So the acquisitor canceled the Wilson shares (47% of the total) of the stockholders who did not accept the offer, and sent them the LTV preferred stock in its place. On a Sunday morning in London, *The Financial Times, January 8, 1967,* reported:

> This is the first time that a Eurodollar loan has been made in Europe specifically, for use in America. The loan, put together by M. C. Rothschild of London and Lehman Brothers of New York will be for the unusually short period of two years.

The period of indebtedness for LTV was to be even shorter. But Wilson & Co.'s expense for its surrender to LTV control is not so short. In order to pay LTV $50 million so LTV could quickly repay the European bankers, Wilson & Co. (as an LTV subsidiary) sold stock and subordinated debentures for which the meat-packer became obligated to pay principal and interest for a quarter-century.[16]

Wilson & Co.'s net worth before merger was $125 million. Transference to it of $50 million of the acquisition debt diminished the company's net worth by 40%. It would be restored by consumers of Wilson products.

But in the words of James Ling, it is "only proper" that the acquired company bear the acquirer's purchase debt.[17] Propriety does not come cheaply to Wilson & Co.—nor to its customers.

Strength Through Debt

LTV management announced to all shareholders that it had "strengthened" Wilson & Co. by "financially innovative actions."[18] If to reduce shareholders' earnings by transferring to them equity and debt obligations amounting to more than a third of a company's net assets without adding one cent to those assets is to strengthen a company, then that statement is plausible.

As it had with stock of other major acquired companies, LTV placed the weakened Wilson shares on the public market while retaining majority control. That policy, "redeployment" (LTV's word), envisioned that the "earning power of these LTV assets will be translated into substantially greater values as measured by the market."[19]

But carrying "redeployment" a step further, for the first time the parent company divided the assets of the acquired company itself into publicly traded entities. Wilson Sporting Goods and Wilson Pharmaceutical Co. were severed from the meat-packing enterprise. Each appeared independently on the American Stock Exchange along with Wilson Meat Packing Co.

By the fall of 1969, however, the market was no longer "translating" LTV-controlled assets into greater values. The earnings of the three Wilson companies combined had fallen more than 40% since the surrender of independent management.[20] But even though LTV initiated no new acquisitions in 1969 (the first year in five it failed to do so), "redeployment" continued—the market would have another chance to reflect the greater value of the assets' earning power. Through the exercise of majority control, LTV took assets from Wilson Meat Packing Co. and divided them into four more publicly traded companies: Wilson Certified Foods Co. (beef, veal, lamb, and pork), Wilson Beef and Lamb Co. (fresh beef and lamb), Wilson-Sinclair Co., and Wilson Laurel Farms (poultry).

The next year LTV sold all its holdings, not just majority control, in the two most profitable Wilson companies—the sporting goods manufacturer and the pharmaceutical company. Ling tersely said that the sale was necessitated by "bank debt which carried with it heavy interest burden owing to prevailing high interest rates."[21]

Devastating Incident

The extremity of having to pay old acquisition debts not with income from the acquired assets, but rather with proceeds from their sale, came after, in Ling's words, the "single most devastating incident in LTV history."

By late 1968 LTV had successfully completed the acquisition of Great America Holding Co. By the standard of asset value, it was double the size of Wilson & Co. Eighty percent ownership of Braniff Airlines constituted only a quarter of the holding company's total value.

Public recognition of the value of LTV's marketed assets was at its zenith. Great America shareholders gave up their ownership rights not for ownership of LTV but, surprisingly, for LTV debt securities and warrants to purchase LTV common shares at a price of $115. Thus, the former Great America shareholders had, for each share exchanged, besides the debentures, the right to purchase for $115 a stock that in late 1972 sold at around $12.

Flushed with the success of having gained $225 million worth of assets, again without having to give up even one share of its ownership (by issuing common stock), LTV set the stage for its most colossal acquisition—Jones & Laughlin Steel Corp. with assets in excess of a billion dollars. But the strategy of acquisition hinged not on the issuance of debt securities as in the Great America purchase. Rather, the takeover of Jones & Laughlin, like the takeover of Wilson & Co., was by outright cash payment.

First, LTV would purchase 63% of the steel producer's stock with $428.5 million largely from bank loans collateralized with securities of earlier LTV acquisitions (Braniff, National Car Rental, and Computer Technology). The Jones & Laughlin shares thus purchased would serve as collateral for further purchase of the same stock.

Second, LTV would create a shell corporation—the J & L Corp.—to hold the purchased Jones & Laughlin stock. Debentures, warrants, and common shares of that holding company would be issued in exchange for 18% more Jones & Laughlin stock. Most important was LTV's plan for further issuance of J & L Corp. debentures in order to retire the multi-million-dollar purchase loan from the banks.

Turn of the Tide

The first step went splendidly. The banks were cooperative enough to enable LTV to pay Jones & Laughlin stockholders 70% above the market price, i.e., $85 for shares then selling on the New York Stock Exchange for $50.

The second step collapsed. The tide was going out. LTV could not transfer to the J & L Corp. the huge bank debt incurred by the generous exchange offer. Ling attributed the failure to that "single most devastating incident"—the Justice Department's filing of an antitrust suit against LTV to prevent the merger (see p. 31). Because LTV's right to acquire Jones & Laughlin was then challenged, the public would not accept J & L Corp.'s security issuance.

LTV had to shoulder the debt itself—just when the earnings of all its subsidiaries were declining. Soon earnings were to disappear completely. Worse, operations in 1969 produced a *loss* of $80 million. Gone was any chance that the acquisitions could pay for themselves. LTV could meet the debt only by selling acquired divisions. The best would go first. The astronomically costly amalgamation for which the acquired companies and their customers will pay for decades was coming unraveled.

CHAPTER 6

Acquire One Company with the Treasury of Another

National General Corp. and Leasco Data Processing Co. did not spend cash for their acquisitions. Rather, they acquired cash. Leasco purchased Reliance Insurance Co. for a value of $400 million in Leasco preferred stock and afterward had $80 million of liquid reserves it did not have before. National General purchased Great American Insurance Co. with paper evidencing over $400 million of debt and, after the purchase, possessed $173 million more in negotiable securities.

Both insurance companies, because of longevity and careful management, had accumulated millions of dollars in redundant capital. Redundant because it was far in excess of the capital resources required by law for the protection of policyholders.

Each acquisitor, using the reverse of the LTV technique of transferring debt to the acquired company, accomplished the same ends of self-enrichment by transferring that excess capital from the target company to itself. Leasco used $38 million of Reliance Insurance Company's capital to finance Leasco operations. National General took $172 million of Great American Insurance Co.'s capital and used $13 million to purchase yet another insurance company.

Can't Understand

On the basis of earnings of the last year of independent operations, Great American Insurance Co. was more than three times as large as its acquirer, National General. Reliance Insurance Co. was more than three times as large as Leasco. Ironically, therefore, neither acquirer could pay the purchase debt to the former insurance company owners without the assets that those former owners gave in exchange for securities evidencing that same purchase debt. Thus, the former owners could have retained their insurance companies and directly voted themselves the same type of securities which they took indirectly from the acquirers. The insurance company owners could have given themselves everything they got from National General and Leasco without going through the sale transaction.

So why did they sell? Eugene V. Klein himself, CEO of National General, in reference to "proxy statements" or literature which corporate managements send to shareholders to explain a merger and solicit their approval, stated:

> I believe that the shareholders are not in a position to understand the information because it is so highly legal and technical. Financial information is so complex that I believe it is beyond the capacity of a small shareholder to understand it...I believe that a better way should be found of informing the shareholders of exactly what is going on.[1]

That's saying the inexorable current of corporate concentration is not the product of corporate owners' better judgment. Klein's point: shareholders never ask why when management says to sell. The proxy statements told each insurance company's shareholders they would exchange their ownership for securities of an acquirer made strong by its scattered activities. In fact, the acquisition schemes were a design for obtaining the insurance company money and circumventing laws for the public's protection.

National General Corp. admitted to being a product of diversification from desperation. In 1961, it was an operator of motion picture theaters on the West Coast. To acquire other companies, in diverse fields, was the only way to avoid bankruptcy. Klein explained to the House Antitrust Subcommittee:

> We would have liked to acquire anything that showed cash flow and profit, be it any kind of business, because we were sort of drowning and we were looking for companies to help us get well financially. It was a day-to-day fight for survival.[2]

The company functions as an investment institution, without a planned acquisition program. It does not hire management and does not know in advance for how long it will retain a controlling interest or any interest at all in its purchased enterprises. After two or three years it may decide to acquire complete ownership or to divest completely.[3]

The owners of the first target companies, Designed Facilities Corp., a mobile home manufacturer, and Mission Pak, Inc., a fruit retailer, accepted National General common stock in exchange. They agreed to merge even though theirs were profitable companies, while their acquirer could not exist without them.

"When we got the wrinkles out of our belly," Klein explained, "and we saw that we had a thriving company…we organized our leisure time and financial services concept." National General sold Designed Facilities and Mission Pak "for the most advantageous price."[4]

Devoid of Ownership

So well did National General sell its concept of diversification to Columbia Savings & Loan shareholders that in August 1964, they (as did Great America Holding Co. shareholders selling to LTV) sold their ownership interest for National General securities that were devoid of ownership rights—i.e., they accepted debentures and warrants allowing them to purchase National General stock at $15 per share. The value of the warrants, however, increased as National General continued to take over the earnings of its acquired companies, as its stock soared.

Pre-merger arrangements by four of the eleven National General directors who owned substantial holdings of Columbia facilitated the transaction. With other Columbia Savings & Loan stockholders, the four National General directors, two of whom were also on the Columbia board, and one of whom, Klein, was president of the acquirer and chairman of the board of the about-to-be-acquired company, put together a block of 39% of Columbia stock for assured sale to National General. By the time both boards and both sets of shareholders entered into merger deliberations, therefore, National General needed only 12% more to achieve outright control (51%). The shareholders needed no convincing of which way to vote to be on the winning side.[5]

In April 1968 Klein received a call from Arthur Carter and Sanford Weill of the brokerage firm Carter, Berlind & Weill. They said they wanted to talk over some investment opportunities—possible further acquisitions for National General. Carter and Weill had more details in an ingenious report, "The Financial Service Holding Company," just prepared by the firm. About three hundred copies were being sent out that week. Send one over, Klein said.[6]

The brokerage firm said the report, prepared by its vice president, Edward Netter, envisioned methods for strengthening the insurance industry. Most intriguing were its ideas for debt-straddled companies to take over resources of successful insurance companies. The brokerage firm had mailed two thousand of the reports to prospective corporate acquirers.

Sanford Weill himself would become CEO of an insurance company, Travelers Insurance Group, and take over CitiCorp to form CitiGroup in 1999. With the secretary of treasury he would engineer repeal of the Glass-Steagall Act to make the takeover lawful retroactively. The act had protected bank depositors' money from risky speculation by keeping depository banks and investment houses/insurance companies separated since the Great Depression (see p. 4).

Laws of various states restrict insurance companies from using their resources to control other businesses. Many states, for example, prohibit an insurance company from investing more than 5% of its resources in another company and from purchasing more than 49% of another company's stock.[7] The objective is to prevent an insurance company from overly concentrating its investment resources, necessary for public policyholders' protection, in so few enterprises as to control them.

But an insurance company may create the shell of another company (called a "holding" company, which the insurer owns—or owns the insurer), transfer assets to that shell, and, in the shell's name, begin functioning in other enterprises. The Netter Report merely explored departures from that loophole. An acquiring company (rather than a shell of the insurer's creation), Netter reasoned, could take over or "hold" an insurance company and then transfer the insurer's surplus capital to itself for its own operations. Or better, according to the report, for more acquisitions.

Throw Good Money After Bad

For instance, the Netter Report to Trans World Airlines suggested that the airline acquire an insurance company and then transfer to it the airline's skyrocketing debt. TWA summarized the report on December 12, 1967:

> Our analysis led us to the conclusion that acquisition of a fire and casualty insurance company would be an outstanding diversification vehicle for TWA's financing requirement: It would provide TWA with an immediate source of readily liquid marketable securities that could be converted to cash and utilized in satisfying TWA's financing requirement... An insurance subsidiary would be permitted to purchase the debt of TWA...and therefore could be used to supplement TWA's other sources of debt capital.[8]

As might be expected, the word for a company in distress to latch onto assets of a healthy enterprise was "diversification." Unlike the report itself, the TWA summary did not bother with the pretense that the acquired insurance company's reward would be anything other than the looting of its treasury.

The prime recommendations of Carter, Berlind & Weill (CBWL) were, naturally, insurance companies that had, through virtue of long and constructive operation, the greatest amount of surplus capital. Listed as the choicest fruit were Great American Insurance Co., Insurance Company of North America, Hartford Fire Insurance Co., and Reliance Insurance Co. "Computation of Insurance Company of North America's capital redundancy," read the report, was $495.5 million as of December 31, 1966.

The Netter Report naturally went to corporations in need of cash. Secure funds of the insurers would be risked to aid sick investments. The brokerage industry's "finest example of creative research,"[9] sought to circumvent laws and throw good money after bad.

Describe as it might the scheme as a godsend to the insurance industry, CBWL could not deceive itself. The brokers well anticipated the reaction of any insurance management to a plan for taking over its treasury. The Netter Report stops short of advice for genteel negotiations. In fact, it actually advises against conversations with the prime candidates' managements. TWA pondered the report:

> In our discussions with Carter, Berlind & Weill they indicated that in their opinion it would not be advisable to attempt to discuss merger with any of the aforementioned fire and casualty companies. Management for the most part, does not appear to be inclined toward the merger route.

Management's revulsion didn't matter. An insurance company's stockholders could just as well be dealt with directly:

> Therefore, it would appear that the most feasible course to pursue would be some sort of a tender offer for at least 51% if we so desire.[10]

No need to purchase more stock than necessary to take control of the treasury.

Surplus Money

However, Netter did have discussions with the management of one candidate. A. Addison Roberts, president of Reliance Insurance

Co., responded with a lawsuit charging that CBWL and others were conspiring to manipulate the price of Reliance stock.[11] But the suit made no impression on Reliance stockholders. They saw no reason to think beyond the description of the benefits of "diversification" set forth in the exchange-offer literature. The acquisition of Reliance Insurance Co. was the initial outstanding success of the Netter plan.

The CBWL brokerage firm denied that the scheme of the Netter Report was to use insurance capital to acquire diverse target companies.[12] But the report under "Leverage" states outright:

> The degree of insurance premium volume written as a percentage of adjusted capital and surplus will determine the potential availability of capital for immediate diversification.

The report does not describe other uses for the surplus capital of successful insurance companies: instead of augmenting the current of corporate concentration by corporate acquisitions, those funds can be used to reduce policyholders' premium payments or ensure them better protection. Policyholders' premium payments, after all, are the original source of the surplus capital.

When preparing suit to prevent then the largest insurance company takeover in history, the Justice Department learned that acquisitors' takeovers of insurance assets had markedly reduced the insurers' capacity to protect the public. But it promptly abandoned the suit (see p. 291).

Eugene Klein read his copy of the Netter Report and decided it had possibilities for National General. He returned the call from Arthur Carter and Sanford Weill and suggested they meet on May 9, 1968, for dinner at Club Twenty-One. He would gladly give them information on National General. The corporation had just finished acquiring Grosset & Dunlap. Klein saw no reason to stop there. Four days after the dinner, Klein called Weill and said he was ready to decide on a definite insurance company target. Weill had carefully kept Klein informed of Great American's stock holdings.[13] The two agreed to aim for Great American Insurance Co.

On May 28, 1968, National General's executive committee authorized the purchase of 500,000 shares of Great American at not more than $55 per share. CBWL said it knew where to locate block holdings. By instigating a merger with its Netter Report, the brokerage firm sought not only a million-dollar finder's fee but also commissions from the transfer of stock—based on the firm's buying and selling transactions—and, later, even fees for acting as agent during the tender offer.[14]

Suddenly a cloud appeared. One day in early June, Klein studied the board tape from the New York Stock Exchange. To his consternation he read that another suitor was bidding for Great American. AMK Corporation was negotiating directly with the management.

A crisis session assembled. To purchase enough shares direct from the stockholders, National General first had to buy a big block of the stock outright. To raise the money to buy it, National General would have to sell its Designed Facilities subsidiary. But it would not receive that money until July. AMK Corporation could complete its transaction in June.

CBWL knew where to purchase 266,300 Great American shares. But try as it might, National General could not come up with the cash. The plan was falling apart just because of a lapse of three weeks.

For CBWL to transact the purchase without possessing the money would amount to a violation of the principal rules of the Federal Reserve Board. Regulation T requires, understandably enough, that a purchaser must pay for the shares he orders within seven days.[15]

But the brokerage firm, which stood to gain millions in finder's, transfer, and exchange fees through implementation of its plan to take advantage of a loophole in the law, was not to be cowed by a Federal Reserve Board regulation that the board itself does not bother to enforce. Entrusting chickens to the fox, the board leaves enforcement of its regulations to the New York Stock Exchange.

Markets Unreasonably Upset

On the morning of June 11, a CBWL broker called Klein and asked, "Will you purchase the 266,300 shares if you cannot get twenty-one day delivery?" Meaning, will you purchase if you have to pay within seven days as required by law?

"No," replied Klein. "I cannot purchase and will not purchase unless I can get twenty-one day delivery [wait twenty-one days to pay]. Now check with your people and see if it can be done."

He had done it many times before, Klein told the broker. A few minutes later, the CBWL broker called again and said, "Confirmed."

Klein: "OK, We have a transaction."[16]

Inadequacy of funds to support stock exchange transactions led to the 1929 stock market crash. Regulation T, which requires adequate payment, is a cardinal reform of the Securities Exchange Act of 1934.

The New York Stock Exchange determined on June 14, 1968, that CBWL had violated Regulation T by arranging for 21twenty-one-day delivery. The exchange, however, did not require that the purchase of Great American shares be rescinded as the regulation prescribes. Rather, it decided that "liquidation of the purchase would unreasonably upset the markets in both National General

Corporation and Great American stock." The exchange thus reasoned, in an opinion of August 15, 1968, that "exceptional circumstances" warranted extension of time for CBWL to complete the transaction:

> It was decided, therefore, that while "exceptional circumstances," justifying an extension of time, did not exist at the time the transaction was executed, the results of the execution had created such "exceptional circumstances." The Exchange then granted an extension.[17]

By that reasoning, the more reckless the violation of Regulation T, the greater the chances the New York Stock Exchange will permit the violation.

The excuse CBWL gave the New York Stock Exchange was that Regulation T is too complicated to be understood. The firm produced "one of the finest examples of creative research ever seen in the [brokerage] business" (its own words). Yet CBWL did not know, its managers explained to the New York Stock Exchange on September 3, 1968, that its delayed payment for the purchase of Great American Insurance Co. shares for National General amounted to a violation of Regulation T:

> As we have previously stated to the Exchange, neither Carter, Berlind & Weill, Inc., nor persons associated with Carter, Berlind & Weill, Inc. were aware that the contract at the time it was made was not in conformity with the requirements of Regulation T. On the contrary, it was believed that the form of contract—seller's option—as an NYSE-approved form of contract (the other approved

forms being Cash, Next Day, Regular Way, etc.) satisfied all appropriate regulatory requirements.[18]

The New York Stock Exchange actually believed that. Ignorance of the law, it replied, is no justification. "In very simple terms," the exchange told CBWL, "a firm that places its services before the entire investment community must be absolutely certain it knows and understands every rule application." On October 15, 1968, the firm's managers read from their teletype that though they received a fine of $5,000, the transaction would stand.[19]

From that unlawful transaction, with the $5,000 fine deducted, CBWL gained $425,975. The exchange's acquiescence to a Federal Reserve Board rule violation hinges, then, on the profitability as well as on the magnitude of the unlawful transaction.[20]

Compliance—A Private Matter

Under the present system of relegation to the New York Stock Exchange of enforcement of Federal Reserve Board regulations, the exchange need not report to any government authority or make public disclosure of violations. Compliance with the government regulations it enforces is thus a private matter between the exchange and its members:

Mr. HARKINS [House Antitrust Subcommittee counsel]. What are the S.E.C.'s normal procedures for review of the stock exchange disciplinary procedures?

Mr. BUDGE [Securities and Exchange Commission Chairman]. Under the decision of Congress the New York Stock Exchange is a self-regulatory body charged with being responsible for the conduct of its membership. The

Commission has an oversight responsibility over the New York Stock Exchange. Normally, there would not be a disciplinary action taken by both against a member firm.

Mr. HARKINS. Are the procedures applicable to this relationship spelled out in a regulation?

Mr. BUDGE. I think not. As a matter of fact, I am not sure that until very recently the Commission was even aware of the disciplinary actions taken by the New York Stock Exchange against member firms. I think historically we have not been advised of that. Is that correct, Mr. Loomis?

Mr. LOOMIS. We were not as well advised as we are now, although we did get a certain amount of information, I think, along.

Mr. HARKINS. How well are you advised now?

Mr. LOOMIS. I think quite well.

Mr. HARKINS. When was this change put into effect?

Mr. LOOMIS. It has been a gradual thing.

Mr. HARKINS. Are there specific regulations requiring submission of anything concerning disciplinary action?

Mr.LOOMIS. There are no specific regulations requiring it. We just get it.

Mr. BUDGE. The person against whom the disciplinary action is taken, of course, has the right to appeal to the Commission. No, I guess that is with the NASD. We would get it if it came through the National Association of Securities Dealers.[21]

Thus, if the New York Stock Exchange does happen to notify the Securities and Exchange Commission of a member's violation of provisions for investor's security, according to the SEC chairman, the exchange so reports because it so condescends.

Now the battle was joined. Of the two suitors, Great American saw rival AMK as the lesser evil. On July 15 the insurance company advised its shareholders against accepting National General's offer of debentures (IOUs) for their common stock. On July 18 it announced, "The management…is giving favorable consideration to a proposal by AMK Corporation."[22]

National General countered on July 29 with its own advice to the same (target insurance company) shareholders: "If the AMK proposal is presented to the stockholders of Great American, as substantial holders of shares we intend to vote against it, and will urge our fellow shareholders also to reject it."[23]

In the end, National General won. For the company that possessed approximately $300 million in surplus capital alone (including that required by law), the acquirer paid approximately $500 million in debt securities. By the time of the hearings at which Klein testified, the value of all the securities the former Great American shareholders took in exchange did not even equal the amount they would have secured had they liquidated that surplus capital rather than selling to National General.[24]

The acquisitor's purchase debt, paradoxically, as already seen, could not be paid to the former Great American owners without the wealth and earnings they gave the acquisitor.

Insurance Treasury Cut in Half

Great American Insurance Co. had never paid a dividend of more than $2.50 per share. On January 14, 1969, during the first year of control by National General (owner of 98% of Great American stock), the insurer paid a dividend to National General of $55 per share. In all, $173 million. The majority shareholder thus shaved the reserves down to the very limit required by law as minimum protection for policyholders. On April 2, 1969, the superintendent of insurance of the state of New York received from his staff an opinion that the extraordinary dividend squandered funds for public safety:

> However, the loss of 57% of policyholders' surplus may inhibit Great American's ability to continue to underwrite its historical proportion of the total demand for insurance coverage and to maintain its competitive position. In such respects, aside from any questions of legality, the dividend action instituted at the first Board meeting after the change in ownership cannot be considered to have been in the best interests of the insurer and the insuring public.[25]

Before acquisition, the insurer had the freedom to greatly expand its underwriting. After National General's declaration to itself of a $173 million dividend from its policyholders' reserves, Great American was bound to its current level of business and could not rise above it. The company had lost "potential for substantial growth," the superintendent's opinion continued:

> It is apparent that the payment of the special dividend of over $170,000,000 not only represents a loss of more than $5,000,000 annually in investment income but sharply

reduces the possibility of material future capital gains, since the securities transferred were the equivalent of one-half of the common stock portfolio. Thus the Company has lost an important potential for substantial growth in the future—to maintain and increase its ability to respond to the growing demand for insurance and still increase its surplus.

The Insurance Department of New York pondered not only whether Great American's operations would be permanently curtailed by the surrender of its capital reserves, but whether that most successful of companies in its field had "commenced voluntary liquidation."[26]

Directors Marching Arm in Arm

"We asked management what was a safe dividend to declare. They came up with a recommendation and figure and we accepted their recommendation and subsequently declared a dividend," the president of National General explained during congressional hearings.[27] National General, of course, controlled that management.

A few days after the declaration of the dividend, the Securities and Exchange Commission reminded National General that it had promised as a condition of its offer for the insurer's stock that "No change is now contemplated in the identity, business…" of the insurance company. The commission asked how it reconciled that statement with appropriation to itself of half the insurer's treasury.[28]

The extraordinary dividend declaration amounted to no change in operations because the management, while still independent, had contemplated the very same distribution, National General said.

Secondly, the commission inquired, what would the acquirer do with its $173 million dividend?

"We are unable to reply definitely at the present time," National General answered. But have no fear, it assured: possible acquisitions were "not yet even conceived."[29]

One week and one day after that reply, National General announced its offer of $15 cash for each share of stock of a second insurer, Republic Indemnity Co. The $13 million cash paid for 97% of Republic's stock came entirely from the Great American cash dividend. National General's management had entered into an agreement to purchase Republic on January 16, two days after the declaration of the Great American dividend.[30]

Five owners of substantial holdings of Republic stock were officers or directors of the acquirer. Symmetrically, the National General board chairman and president, Eugene Klein, was a Republic director; the Republic board chairman, Marvin Finell, was a National General director. Together, the two board chairmen owned over a third of Republic's voting stock. Finell was also chairman of the Great American operating committee. Several days before the declaration of the $173 million dividend, he had received a fee from National General for his legal advice on "whether to go forward on Great American...and other items involving that corporation."[31]

National General's management maintained that $15 per Republic share was a fair price to pay, that the amount would have been lower if the purchaser had not been bidding against another contender. Nevertheless, those five officials, while deciding the amount of National General's offer, were in effect determining the portion of the Great American dividend that would be paid to themselves. Together they received for their personal accounts over half the amount paid for Republic Insurance Company.[32]

By January 1970 the fate that the Netter Report plotted for the ten wealthy insurance companies had befallen eight of them. Klein explained to the House Antitrust Subcommittee that the wealth of those companies was a sign of poor management and, for that reason, of their need to be relieved of their independence. The subcommittee chairman, Emanuel Celler, was curious:

> In other words, this document prepared by the firm Carter, Berlind is sort of an inducement for companies like your own to make an acquisition of an insurance company because it has in its portfolio a great many stocks and bonds the total value of which is beyond that required by the authorities. A so-called surplus is required to take care of contingencies by floods, hurricanes, what-have-you, whatever you may call it. Surplus-surplus is the excess over that amount. This Carter, Berlind document encouraged you, and others like you, to take over such a company; and when you take over that company you just siphon off for your own purposes this so-called surplus surplus.
>
> Now, don't you think that when an insurance company has been in existence for many years, and acquires a portfolio of various types of investments—in your case they were blue-chip stocks, I understand—that that fund cannot be looked upon as an ordinary surplus, and isn't it in the nature of a trust fund for the policy-holders?
>
> Do you think that you, as a principal stockholder, or your colleagues as principal stockholders, have an inherent right to shave off that surplus surplus and use it for your own purposes?

Mr. KLEIN. Yes, sir.

The CHAIRMAN. Isn't an insurance company different from an ordinary business? If you acquired an ordinary business operation that had a tremendous book value, a tremendous portfolio of finer stocks the cash value of which wasn't needed in the operation of the business, I would say, "All right, take those blue-chip stocks and divide them among the stockholders."

But here you have an insurance company built with blood, sweat, and tears over the years. It is an old company you took over, and it had this surplus. You come along, and I can use the word "ruthlessly," take over this surplus surplus—even if you had the consent of the superintendent of insurance of the State of New York. I think there is something amiss here and should have given pause...

Mr. KLEIN. I disagree with the contention that we did anything wrong: I mean specifically that we were ruthless. That is a tough word, Mr. Chairman.

The CHAIRMAN. I am meaning to be tough.

Mr. KLEIN. I understand that, sir. No. 1, at no time and under no circumstances did the declaration of that dividend do any harm to the reserves of the policyholders of the insurance company. The remaining reserves were totally adequate, more than adequate to meet any contingency, any emergency, any disaster, any flood, as proven by the fact that 1969 was one of the worst years the insurance company ever had in all its existence, and incorporated into 1969 were such disasters as Hurricane Camille, which cost the company in excess of $6 million.

So, in the worst year, or approximately the worst year of the company's history, with untold disasters and catastrophes, such as Hurricane Camille, the company met its obligations with no strain whatsoever.

This company...is over a hundred years old. These securities have appreciated greatly. This is true of many insurance companies, and it is also true that, because of that very strength, that strength that was not needed, it had more than twice as much financial strength than was needed by any stretch of any imagination; it had, in my opinion, and our opinion due to a great deal of laxity on the part of management of insurance companies.

The CHAIRMAN. Will you yield a moment?

Mr. KLEIN. Yes, sir.

The CHAIRMAN. Why couldn't you have, in the public spirit and in the interest of that company and public relations, spread the good gospel of your company by taking that huge surplus and, by the use of it, reduce the rates to your policyholders? Why didn't you do that?

Mr. KLEIN. Well, that is a different question, sir, and I will come to that.

The CHAIRMAN. It is quite relevant.

Mr. KLEIN. No. 1, I believe that an insurance company, like any other business enterprise in the American system of business, seeks to operate at a profit.

The CHAIRMAN. Are you an expert in the insurance business?

Mr. KLEIN. No, sir.

The CHAIRMAN. How long have you been in the insurance business?

Mr. KLEIN. I was a director of an insurance company for approximately 5 years before.

The CHAIRMAN. Only 5 years and you have been in an operating position now for a year, is that right?

Mr. KLEIN. Correct.

The CHAIRMAN. Or rather a year and a half. Do you deem yourself an expert on insurance?

Mr. KLEIN. No, sir.[33]

Large Blocks Located

In the fall of 1967, the Netter Report fell on the desk of Saul Steinberg, 28-year-old chief executive of Leasco Data Processing Co. Leasco had been incorporated in 1965, and in two years its income, mainly from leasing computers, had risen from $196,000 to $1,389,000. The investment community regarded the company highly, awarding it a price of thirty times earnings, or $62 per share.

Of the report's prime candidates, Steinberg chose Reliance Indemnity Co. The redundant surplus capital—the amount of capital reserves in excess of that required by law for policyholders' protection—by the most conservative estimate surpassed $80 million.

In "A Confidential Analysis of a Fire and Casualty Company" of January 11, 1968, Leasco said that capital was the apple of its eye. A merger with Reliance would automatically increase Leasco's per-share earnings because Reliance's shares sold at a lower price in comparison to earnings.[34]

Steinberg and three other Leasco officials met Netter and other CBWL members at the brokerage firm's office, also on January 11, 1968. One Leasco representative was five minutes late and thereby missed the entire meeting. "We were annoyed and we got up and walked out," Steinberg stated. As soon as the Leasco officials had walked in the door, CBWL had handed them a memorandum. Listed were three demands. First, "Carter, Berlind & Weill will be given one seat on the Leasco Board immediately and acquired company board subsequently." Second, "All purchases of stock in the acquired company on or off the New York Stock Exchange will be made exclusively through Carter, Berlind & Weill." Also, the firm "will have a joint managerial position on future public offerings and private placements." Third, the firm demanded a $750,000 finder's fee.

That list "really jolted us," Steinberg said.

> We left. We laughed. We were weak. It really jolted us, and we didn't expect that our future would be edicted to us. We showed him [Leasco's investment banker] the list of demands and he said in his forty years of Wall Street he had not encountered such a thing.[35]

But in spite of that awkward beginning, by spring bygones were bygones. CBWL modified its demands somewhat, Steinberg met

Netter, Weill, and Carter at Club Harmonie for lunch, and both parties agreed to cooperate for the common goal.

The brokerage firm's cooperation was essential to Leasco because the brokers knew where to locate the large blocks of Reliance stock. With particular savvy, the firm knew which mutual funds and other institutions to approach.

As the season wore on, however, Carter, Berlind & Weill grew impatient. The firm thought Leasco was dragging its heels, and, worse, was not laying all its cards on the table. On June 11, 1968, CBWL brokers urged Steinberg to announce a tender offer for Reliance stock the very next day. "How can I talk with the funds," Carter asked, "if you don't give me any information? You are not keeping me informed."[36] He urged Steinberg to offer $120 per Reliance share.

"Astonished, we [Leasco] said that with that view he could not possibly have analyzed the economic consequences of such a transaction." On second thought, the CBWL brokers suggested the price of $85 per share. Carter argued on behalf of his firm's customers—a group that included the firm's members—stating that they would have to receive substantially more than they had paid for Reliance stock.[37]

Six months earlier, in December 1967, Carter, Berlind & Weill had urged Reliance Insurance Co. to sell out for $45 (in paper, not cash) per Reliance share—approximately half the price it was now urging Leasco to pay.

The brokers' impatience with Steinberg was nothing compared to their exasperation with A. Addison Roberts, president of Reliance. To the firm's dismay, Roberts was content to manage the insurance company. Obviously, he represented the competent management that turns a company into a broker's target.

The discontent worked both ways. CBWL's circulation of rumors of an imminent takeover of Reliance did not sit well with

Roberts. Netter was attempting to drive up the price of Reliance stock by stating that Carter, Berlind & Weill were the managers of the merger when as yet there was not even a merger partner. According to Roberts:

> For a long time we knew that Carter, Berlind [CBWL] were stating to people that they were definitely going to involve us in a deal. But…[said those people] they [CBWL] wouldn't tell us who it [the would-be acquirer] is. It was a misstatement to buy this stock. This is a good opportunity to make a profit.[38]

Can't you show a little interest, to give some substance to CBWL's rumors, Netter said to the president of Reliance. "Won't you say to me that you're going to make a deal?" he pleaded.[39]

No Cooperation

Roberts recounted: "He also, as you might know, called my secretary and asked her what my reaction had been to the Eaton, Yale & Towne suggestion and that time was running out on me if I didn't do more."[40] Netter had in mind the forklift manufacturer Eaton, Yale & Towne and Gulf & Western, besides Leasco, as possible Reliance merger partners.

If only Roberts would cooperate, CBWL told him in February 1968, the firm would put Reliance stock at $40 or $50 per share—the stock for which in June it would ask $125.[41] But CBWL's efforts were succeeding without Roberts' cooperation. By May Reliance stock had risen to the point that the insurer's board of directors feared that the shareholders could be tempted to sell. The directors thought, the president recounted, "we were going to be forced into a deal and I was authorized to find any company that I thought would be acceptable for us."[42]

Did he then discuss merger with many other companies in an effort to supplant Leasco with an acquirer of Reliance's own choice? Yes. With "half of America it seemed like."

Leasco purchased $4,400,000 worth of Reliance stock in March for prices between $30 and $40 per share. In April the stock rose above $50, causing Steinberg to consider changing the target to the lower-priced Hartford Fire & Casualty Co. (not yet acquired by International Telephone & Telegraph Co.).

But on June 22, 1968, to CBWL's delight, Leasco at last issued a press release announcing its intent to tender an offer to the shareholders.[43] A month later, on July 23, Leasco offered Reliance owners $72 in Leasco securities for each share.

Both Buyer and Seller Be

Reliance management, in its own words, was "totally unreceptive." On the same date, it informed its shareholders by letter that Leasco's entire earnings for the past five years equaled little more than half the $9 million to be paid CBWL for exchange transaction expenses alone.[44]

Reliance management advised its shareholders that those brokerage expenses were exorbitant: "We caution you that brokers' opinions favoring merger may be influenced by the generous fees to be paid them by Leasco." CBWL received almost a quarter-million dollars in fees just for transfer of the shares it possessed. During the first half of 1968, CBWL was encouraging its customers to buy those shares from which purchase it would receive a commission; at the same time it was arranging their sale to Leasco, from whom it would receive another transaction fee. It was arranging buy and sell transactions of the same securities at the same time. Inexplicably, CBWL stated to Senate investigators that it had no part in the plan for resale of those shares to any acquiring company.[45] At the congressional hearings, A. Addison Roberts, still president of one-hundred-year-old Reliance, sat beside his new guardian superior, Saul P. Steinberg, chief executive of Leasco. As head of Reliance, he said, "I thought best for Reliance to stay independent." He would still be resisting the takeover by suing Leasco and CBWL were there any chance:

Chairman CELLER. What happened to the lawsuit, Mr. Roberts?

Mr. ROBERTS. On the advice of our lawyers we withdrew the lawsuit because they felt that we could not prove the conspiracy.

Chairman CELLER. Or was it due to the fact that a better offer was made subsequently?

Mr. ROBERTS. No, sir. Let me say this: If we had thought this could have been won, I think we still would have been fighting.[46]

With the $400 million acquisition of the enterprise that surpassed Leasco's earnings ten times over, the acquirer's net worth increased 1,361 percent, from $16 million to $236 million. Net income increased 1,834 percent, from $1.4 million to $27 million.[47]

Metamorphosis

Admiration for the Leasco-National General concept of taking gross dividends from acquired treasuries, and for the LTV concept of loading debt on the acquired company in order to pay for it, has led to a metamorphosis of those two devices. Private equity firms may take over companies that do not have rich treasuries. They nevertheless pay themselves mega dividends from the takeovers by loading them with debt in LTV style.

Blackstone Group bought Celanese Corp. for $3.4 billion in 2004. After paying less than a fifth of that purchase amount from its own funds, it created new debt for Celanese. Celanese then paid Blackstone a $1.3 billion dividend.

Kohlberg Kravis Roberts & Co. acquired PanAmSat in 2004 for $4.3 billion. A month later PanAmSat issued $250 million in notes to pay a dividend to its acquirer. "The huge fee and dividend boom may not be sustainable. Market conditions are deteriorating. Interest rates [target companies must pay to issue dividends to their acquirers] are going up," *The Wall Street Journal,* January 5, 2006, reported.

"Celebrated buy-out firms like Blackstone Group and Kohlberg Kravis Roberts & Co., hailed only a year ago for their deal-making prowess, are seeing their profits collapse as the credit crisis spreads," *The New York Times,* March 11, 2008, reported.

CHAPTER 7

Accept Delivery from the Pentagon

In the late spring of 1966, George Manis, chairman and principal stockholder of Memcor, Inc., called Chief Executive James Ling to ask if Ling-Temco-Vought, Inc., would be interested in merger. Manis said he was in poor health and could no longer manage the company. Memcor produced electronic equipment, principally the PRC-25 walkie-talkies used by the U. S. Army in Vietnam and by foreign military services.

Memcor's drawing card, Ling's subsequent investigation report ironically disclosed, was the company's imminent collapse. It could not fill large backlogs of walkie-talkie orders. To obtain those orders, it had underbid RCA by more than $200 per instrument. The undated report read:

PROGRAM STATUS

1. Behind on all deliveries...has delivered about 7,000 units and is about 10,000 units behind.
2. Ruffin [president of Memcor] was able to con Philadelphia into a 95% progress payment under (1) above. They have used up 90% of the 95% to date.
3. RCA's present price on the PRC-25 is $820 per unit; MEMCOR's bid was $615 per unit at one time. Looks

as if MEMCOR bought into program with hopes of re-couping by the ECP route, which did not happen.

4. Even though MEMCOR is very late in delivery...the Army will not press for liquidated damages since the equipment is urgently needed for Southeast Asia.[1]

To have "bought into the program in the hopes of recouping by the ECP route" means that Memcor had recklessly submitted a lower-than-possible bid to remove other producers from the competition. The only hope of salvation after that desperate gamble would be the Pentagon's granting of an Engineering Change Proposal—a euphemism for an agreement to pay Memcor more than the price arrived at by competitive bidding for which Memcor promised to perform the contract.

The gift of that price increase from the Pentagon would mean that the system of competitive bidding—by which only the most competent producer should win the contract—would be thrown to the winds. Although LTV considered that Memcor was "'in' with the Army," the acquisitor concluded that there was no chance that the Pentagon would free Memcor from its own trap. It could not envision such a gratuity as a Pentagon price increase that would reward a contractor for scheming to eliminate competition. A memorandum of June 29, 1966, to LTV Vice President J. W. Dixon, based on confidential information from the Army Electronics Command, stated:

Memcor equipment is definitely good, but it is late. The company is losing money on each one because their costs are too high and their costs have gone up since they bid the job. They are trying to get contract modifications to bail them out, but they are not going to get them through.[2]

LTV guessed wrong. The Pentagon did grant Memcor's requested increase.

LTV Vice President Dixon concluded, after an inspection trip of to Memcor facilities in Huntington, Indiana, and Salt Lake City, that various competitors stood ready to produce the product that Memcor could not "deliver." He wrote to Ling and LTV President Clyde Skeen on July 15, 1966:

> Customer Relations and Competition: Because of inability to deliver, resulting from financial problems, customer relations are anything but good. The army has done everything possible to help Memcor, including increasing progress payments on PRC-25s to 95% but will not give them more business unless they demonstrate performance. Additional PRC-25 spares business is currently going to RCA and other companies on a component basis. The Army has indicated they do not think highly of Memcor's present operating management.[3]

Dixon's report specified the other companies capable of producing the PRC-25:

> In the components area, Memcor said the following companies were their prime competitors in an annual market of $75 to $100 million. These are: Ohmite, Ward-Lenard, Sprague, Mallory, and International Resistance Corporation. The latter three of these companies are growing and profitable organizations.

Memcor required fifty man-hours to produce each PRC-25 unit. LTV's information was, however: "Attainable man-hours: 30. RCA is doing it in 32."

Revenues Mingled

Another company apparently capable of PRC-25 production was Magnavox, the sales agent for all Memcor products in Europe. LTV inspectors reported to Ling on July 14:

> The Swedes came to Magnavox and insisted they [Magnavox] be the export agent for their order of the PRC-25's to Memcor. The Swedes have released $2.5 million to a United States bank, but the bank has to guarantee that this money will not be spent on anything but the Swedish order. If Memcor cannot deliver the PRC-25's to the Swedes, they are going to turn all drawings, inventory, work-in-progress, etc. over to Magnavox, and Magnavox is to complete the order.[4]

As with the Wilson merger, a major consideration was the effect that the "pooling" or mingling of Memcor revenue from defense contracts with LTV's revenues would have on LTV's earnings reports. For again, the plan was to give in exchange the acquisitor's securities earning less profit (in proportion to their market price) than were the securities to be acquired. This time, pooling of revenues would produce results more spectacular. On July 18, LTV finance manager George Griffin wrote to Ling and Skeen:

> Assuming 600,000 shares LTV Electrosystems Inc. [the acquiring LTV subsidiary] common are exchanged for Memcor

shares, Memcor would increase LTV's [Electrosystem's] 1966 earnings by about 10-15% (including 9 months of earnings last quarter FY 1966 at break-even).

But if LTV Electrosystems preferred stock were given for Memcor stock, through pooling-of-interest accounting, the LTV subsidiary's earnings could increase by as much as 25 percent. And if debentures, by 50 percent.

Assuming a 5 1/2 % convertible preferred at $9 million Memcor would increase LTE's 1966 earnings by 20-25% on the same basis. If a convertible debenture were used, there would be an increase of twice this rate, because of the deductibility of the interest. However, the use of debentures would make the transaction taxable and therefore, less desirable to Memcor stockholders.[5]

The sharp rise in earnings would increase the value of LTV stock and yield greater wealth to take over more corporations.

When asked whether to increase LTV earnings was the real purpose of the acquisition, Skeen said, "First was to get into an area of electronic products that none of the LTV companies was in, and secondly, obviously, to increase the earnings of the Company if we can make it profitable."[6]

On July 21, 1966, Ling recommended to the Board board of Directors directors that LTV purchase Memcor:

Incidentally, Memcor is not shopping for a merger or to be acquired but because of Mr. Manis' serious illness for over a year, coupled with his feeling that Memcor's long-term outlook would be better teamed with LTV Electrosystems, he approached us to explore the feasibility or possibility of merging with Electrosystems.[7]

The boards of directors of both companies voted for merger. On July 25, 1966, a joint press release announced agreement in principle to merge. But the agreement was short-lived. Further LTV investigations indicated more strongly that Memcor could not produce its backlog of orders. Without that production, Memcor's revenue could not heighten LTV's earnings as the finance director had calculated.

In Bad Shape

"It was the unanimous conclusion," Clyde Skeen explained to the subcommittee, "that Memcor was in substantially worse shape than we had anticipated, primarily because their backlog of PRC-25 radios, which was one of their great assets in our view, had a built-in loss of several millions of dollars...Merger negotiations were called off finally and formally."[8]

On July 29, 1966, Ling wired all LTV directors and credit line banks: "Negotiations for the merger of Memcor into LTV Electrosystems were terminated upon mutual agreement of both companies."[9]

Neither he nor the bankers had any idea that a third party was waiting in the wings to make Memcor much more profitable—so profitable that to mingle its revenues with LTV revenue would make the acquisitor's fondest dreams of the paper earnings increase come true. On September 2, 1966, however, Treasurer B. L. Brown returned to the LTV Tower to find on his desk a note from R. L. Thomas, an LTV corporate counsel:

George Bergland (First National Bank in Chicago) called for you today and in your absence, informed me as to the following:

1. The Army wanted us to know that it had increased its quantity order on the three PRC-25 contracts by 3.5 million dollars; that it had granted and settled a claim in the amount of $167,000 in Memcor's favor; and that it had removed the liquidated damage clause from the three PRC-25 contracts.

2. The bank wanted us to know that its credit agreement with Memcor had been amended to eliminate any default prior to March 31, 1966, and increased the amount to 3.5 million; that it had extended the equity date to November 30; and that Alexander Grant is now in a position to issue a "clean" audit. All the above, just in case we are still interested.[10]

Sins Funded

LTV finance officer George Griffin quickly called First National Bank to learn more. "Most significant," he wrote to Ling on September 9, 1966, "is the Army's specific request that LTV be informed of the remedial action being taken by them."

"In short," he said, "the Army appears to have done all that might be possible to make Memcor marketable by funding its past sins and anticipated underpricing on the PRC-25."[11]

Not only was the army increasing the walkie-talkie contracts by 25%. It was suddenly recognizing an old Memcor claim—the validity of which not even Memcor's auditor discerned. Griffin continued:

Tyler Port, Under Secretary of the Army, specifically wanted LTV to know that the Army had increased the price on the PRC-25 contracts by $100 per unit for the express purpose of making Memcor more attractive to LTV or other parties who could come in and give Memcor the management it needs.

The Army also wanted LTV to know that it had agreed to a settlement in full in Memcor's favor of a pending ASBCA claim in the amount of $170,000. (This also significant to Alexander Grant's opinion, as they questioned the carrying value of this claim.) In addition, the Army has eliminated the liquidated damages clause from the PRC-25 contracts.

Attempting to justify the price increase, the Army Contract Adjustment Board claimed that Memcor's walkie-talkie production was needed in Vietnam, and that without those financial gratuities Memcor could not survive. But the army's decision made no mention of the other companies that, according to LTV's earlier investigations, were clearly capable of the same production.

By transferring the order to those companies, the army could have obtained the PRC-25 production without rewarding Memcor's scheme to "buy into the contract." According to that scheme, as already seen, the company purposely submitted impossibly low bids to eliminate competitors from consideration. Thus, Memcor accurately anticipated that the Pentagon would increase the contract price even though that price surpassed the bids of the eliminated competitors.

If LTV could see so clearly Memcor's "past sins and anticipated underpricing on the PRC-25" walkie-talkie contracts to eliminate competition—with bids so low that Memcor had to have "additional incentive" even to complete the contracts—surely the Pentagon could also.

LTV President Clyde Skeen was asked whether the Pentagon, by raising the contract prices, sought to enhance Memcor's profitability so that it would be an attractive acquisition target:

> Mr. HARKINS. Was the purpose of the Army's contractual actions which have been discussed to assist Memcor in merger?

> Mr. Skeen. Yes, undoubtedly.[12]

Spurious Bid—Real Profit

Thus, the army's justification for instigating the LTV-Memcor merger by rewarding Memcor's spurious contract bid was that the acquisitor would in some way inject financial strength into its acquisition. Two years after the merger, however, LTV did not have the revenue to pay the interest on its acquisition loans, much less to help acquired subsidiaries. With the exception of the first full calendar year after merger, 1968, LTV operated at a deficit. The government's rejection of the competitive bidding process, therefore, produced no sure method for strengthening government contractors.

The Pentagon's $3.5 million, 25% price increase for future Memcor production, the Pentagon's grant of theretofore unrecognized Memcor claims against itself, and the elimination of grounds for its claims against Memcor amounted to delivery of the company to LTV. For then the acquisitor was assured that the profit from the walkie-talkie production would more than offset the amount paid for Memcor and produce the calculated good results on LTV earnings.

To receive delivery without asking causes a person to ponder what he might get if he asks. LTV officials decided to try to procure two million more dollars from the Pentagon by bringing "pressure" on the army to bring "pressure" on the navy to award a Memcor claim that—according to the judgment of LTV officials themselves—"does not appear to be valid."

LTV also planned for the Pentagon to bring "pressure" on Memcor itself to accept a purchase offer favorable to LTV. Assistant comptroller of LTV Electrosystems, B. B. Pettigrew, advised the acquisition managers on November 11, 1966:

ACTION NEEDED

1. Develop an offer.

2. Discuss with the Army (General Latta and Tyler Port) the offer before discussions with Memcor...

The Army wants out in the worst way and will pay to get out and will force Mr. Manis to accept a reasonable offer.[13]

Four days later he advised:

Pressure Navy into paying part or all the $800,000 claim on the URN-20. The claim does not appear to be valid (but it may be a good vehicle), and the Army might have to furnish the funds for its payment (by MIPR).[14]

The Memcor stockholders received $7 in LTV Electrosystems common shares for each Memcor share. As a result of "buying into the contract" and thwarting the competitive bidding process, Memcor thus secured for itself an extraordinary exchange value.

After the merger Memcor profits increased even though sales declined. In granting the price increases, the government did not take half measures. Memcor stands alone as the LTV subsidiary that experienced increased earnings under LTV control.

PART III

✡ ✡ ✡

The Purpose of the Acquisitor's Headquarters

Diversified corporate acquirers rarely if ever reveal whether their management benefits or impairs their acquired corporations. Figures for comparing post-merger performance with pre-merger performance of the formerly independent companies are the acquisitors' most tightly guarded secrets. Acquiring managers, seldom at a loss for words claiming that they improve their acquired companies, do not divulge statistics for proving the point. Describing that reticence of the "conglomerators," Dr. Willard F. Mueller, director of the Federal Trade Commission's Economic Report on Corporate Mergers, said to the Senate Judiciary Committee:

So I am fearful that many people that are doing research on this subject today, who are interviewing leading conglomerators and others who are all charming, articulate people, are going to have a very difficult time cutting through all the jargon and rhetoric to find out what the facts really are.[1]

For want of more revealing disclosure, the financial media often compares the performance of acquisitors' acquired operations with the performance of certain industries. But the comparison offers only vague insight into the question of whether acquired companies benefit from surrender of their independence.

Forbes, January 1, 1973 (p. 156), lists the return on total capital (assets owned outright plus borrowed assets) of forty-six acquisitors. That group's median annual earnings during the five years ending January 1, 1973, were 7.8% of total capital. *Forbes*'s twenty-nine industrial groups—among which the acquisitor group ranked twentieth—returned median earnings of 8.5% of total capital. Thus, the acquisitors' return from assets under their control was less than that of the companies of other classifications. Further comparisons of acquisitors' averages with industrial averages show similar ratios of return on total capital.

The House Antitrust Subcommittee obtained pre-merger and post-merger performance statistics of companies acquired by the four largest subjects of its investigation: International Telephone & Telegraph, Litton Industries, Ling-Temco-Vought, and Gulf & Western.[2] Acquisitors had never before disclosed data for directly evaluating their ability to operate the companies they acquire.

Data for estimating efficiency of the acquiring managements are presented as three ratios: net income as a percentage of sales, net income as a percentage of assets, and sales as a percentage of assets.

In most instances, the last full year of independent management is compared with the subsequent years of the acquisitor's management through 1969.

For example, the data for Fitchburg Paper Company, acquired by Litton Industries, indicate that the rate of net income from sales, 3.4% during the last full year before acquisition, rose to higher percentages during four of the six post-merger years. Fitchburg's rate of net income from assets for all post-merger years except one, however, is lower than for the last year of independent management.

Data are available for twenty-eight of the four acquisitors' acquired companies. Twenty-one of those companies show declining performance as measured by at least two of the three efficiency ratios. Twenty-two of the twenty-eight acquisitions show declining performance as measured by the amount of income generated by assets (see *Hearings on Conglomerate Corporations,* Part 3, p. 191).

Factors besides management efficiency bear on profitability. But proof that at least three-fourths of the acquired companies for which statistics can be produced experienced declining performance after merger casts doubt on acquisitors' vaunted claims that they enhance the operations of their acquired parts. The Antitrust Subcommittee Report states: "Inasmuch as these companies perform in many different industries, and in the light of the fact that until 1970 the trend of annual corporate profits was up, it would be reasonable to conclude that these ratios reflect ineffective management."[3]

Absent Management

So what is the purpose of the acquisitor's headquarters? the House Armed Services Committee asked admirals appearing before it on April 17, 1972. They did not extol "the organization structure we now have to deal with" after takeover of a shipbuilder. The deputy chief of Naval Material for Procurement and Production, Rear Admiral Rowland G. Freeman, said the corporate acquirer imposes a distant "absentee management over the shipbuilder"—a management motivated more by cash than efficiency.

Admiral Freeman flashed on the screen a diagram of the contractor's pre-merger independent organization. He extolled its simplicity and efficiency.

PRIOR TO MERGER

COMPANY A

DIVISION X   DIVISION Y   SHIPBUILDING
                   DIVISION

Admiral FREEMAN. This is an illustration of a ship-building company we dealt with about five years ago. We dealt with them day to day and reached the president easily. The project manager was pretty much in control.[4]

Then the admiral tried to show the diagram of the organization under the acquisitor's control, a labyrinth of offices and divisions. It was blurred and did not fit on the screen.

Admiral FREEMAN. Subsequently this company was bought and merged and this is the organizational structure we now have to deal with. It has high cash motivation, and it's very difficult to reach the people who actually have to make the corporate decision when you are talking dollar schedule performance. It can take months.

His point clear enough, the admiral continued:

There are several problems involved with a large organization [the acquisitor] which is primarily motivated through cash flow and profit, which is what they are in business for—to move money. They are almost an absentee management, or are a large heavily divisionalized management. It is very difficult to get the senior members of the corporation's attention in many cases...We do not get the attention we would like to have.[5]

The riddle of the absentee management remained a mystery.

CHAPTER 8

The Purpose of International Telephone & Telegraph—Captive Customers

After World War II, International Telephone & Telegraph Corp. decided it had set off in the wrong direction when founded in 1920. The company was in the wrong business, it concluded.

While other American enterprises busily prospered from the postwar boom, International Telephone & Telegraph reeled from war damage to its overseas telephone systems. Almost all its stockholders were American citizens, but almost all its assets were in foreign countries. Furthermore, the American Telephone & Telegraph monopoly prevented transfer of operations to the United States. Harold S. Geneen, president and chairman of the corporation, explained to the House Antitrust Subcommittee:

> During World War II, practically all of ITT's companies in Europe were overrun and seized. Currency conversion problems which persisted for many years after the War, and the necessity for rehabilitating and rebuilding our companies, were ITT's primary difficulties. I might point out that these serious losses occurred at a time when most U. S. companies were rapidly expanding from the U. S. plants which had been developed for war time purposes.

ITT had no such advantages. In fact, many of our companies have not been recovered from the World War II expropriations.

Experiences of this kind... convinced the management of ITT of the necessity to diversify into the relatively secure and growing U. S. and Canadian economies. Our first point of entry into the domestic market might logically have appeared to be the telephone equipment manufacturing business, as this was our major field abroad. However, ITT's traditional areas of telecommunications were essentially foreclosed in the U. S. because of the captive market position and vertical integration of AT&T...[1]

ITT, like National General, is the product of diversification from desperation. Yet, if an acquisitor has an authentic function, a benefit to provide acquired enterprises, ITT should be that company.

When acquisitors' stock fell from dazzling heights in 1969 and 1970, ITT conspicuously maintained its equilibrium. By mid-1971, when the wildest imaginations could not envision the return of giant acquisitors' stock even to the low of 1968, ITT easily surpassed its high of that bull-market year.

Its return from acquired assets, however, is not as impressive as its stock market performance. The company ranks twelfth among the forty-six "conglomerates" that *Forbes* lists according to their rate of return from total capital for the five years ending January 1, 1973. Among the entire listing of 780 companies, ITT ranks 274th.

Acquire to Borrow

ITT shrewdly shuns the hackneyed argument that acquisitors deliver financial resources to acquired subsidiaries. The income of a parent acquirer comes only from a subsidiary's treasury or operations. When an acquirer claims that it reverses that process and money flows from it to its subsidiaries, something is wrong. At best, it's referring to shifting of capital from one acquired company to another.

ITT's motives for purchasing Pennsylvania Glass Sand Corp. and Jasper-Blackburn Co. (an electronics equipment manufacturer) dispel any idea that the purpose of acquisition is to bestow, rather than gain, financial strength. Just the reverse is true. ITT calculated before those mergers how much the largely debt-free companies would add to the parent company's borrowing capacity. It determined the amount of financial resources the target companies could provide the acquisitor.

The Jasper-Blackburn Co. of St. Louis, an ITT finance officer informed President Geneen on June 13, 1967, could add $6.5 million to the acquisitor's cash borrowings:

> Blackburn has a strong balance sheet with a current ratio of 5.5 to 1 and over $9 million in equity. The acquisition of Blackburn would generate approximately $6.5 million additional borrowing power for ITT.[2]

The Pennsylvania Glass Sand Corp., President Geneen told the ITT Board board of Directors directors on February 14, 1968, could provide the acquisitor with $28 million borrowed cash:

> Over the past ten years, profit margins have exceeded 15% in every year and have recently increased to a current rate in excess of 20% of sales. The company has operated virtually debt-free during this period, and consolidation of Pennsylvania's balance sheet would add over $28 million to total ITT borrowing power.[3]

When asked whether it was usual to consider how the acquired assets would add to ITT's borrowing capacity, Geneen replied, "There is no question about the point."[4] Table 5 shows how the acquisitor filled its own coffers by adding debt to its takeovers.

Table 5

Selected Financial Stability Ratios of ITT's Major Acquired Subsidiaries[1]
for the Last Audited Fiscal Year Ended Preceding Acquisition
Compared with ITT Ratios for the Next
Fiscal Year Ended after Acquisition

Subsidiary	Long-term debt plus preferred stock to common-stock equity		Long-term debt to net worth		Interest and pre-ferred dividend coverage		Interest coverage	
	Before	After	Before	After	Before	After	Before	After
Avis, Inc	0.52 to 1	0.84 to 1	0.52 to 1	0.58 to 1	10.0 to 1	3.2 to 1	10.0 to 1	3.7 to 1
Airport Parking Corp. of America	0.85 to 1	0.75 to 1	0.85 to 1	0.53 to 1	7.0 to 1	3.1 to 1	7.0 to 1	3.5 to 1
Barton Instrument Corp	0.13 to 1	0.66 to 1	0.13 to 1	0.47 to 1	58.0 to 1	4.6 to 1	58.0 to 1	5.3 to 1
Continental Baking Co	0.27 to 1	1.02 to 1	0.14 to 1	0.56 to 1	10.9 to 1	3.2 to 1	25.8 to 1	4.5 to 1
Howard W. Sams & Co., Inc	0.40 to 1	1.17 to 1	0.40 to 1	0.65 to 1	14.8 to 1	3.0 to 1	14.8 to 1	3.4 to 1
Jasper-Blackburn Co	0.02 to 1	1.0 to 1	0.02 to 1	0.56 to 1	74.6 to 1	3.2 to 1	74.6 to 1	4.5 to 1
Levitt & Sons, Inc.	1.5 to 1	1.0 to 1	1.5 to 1	0.56 to 1	1.5 to 1	3.2 to 1	1.5 to 1	4.5 to 1
Pennsylvania Glass Sand Corp.	0 to 1	1.0 to 1	0 to 1	0.56 to 1	71.7 to 1	3.2 to 1	71.7 to 1	4.5 to 1
Rayonier, Inc.	0.58 to 1	1.0 to 1	0.58 to 1	0.56 to 1	9.2 to 1	3.2 to 1	9.2 to 1	4.5 to 1
Sheraton Corp. of America	3.01 to 1	1.0 to 1	2.9 to 1	0.56 to 1	1.8 to 1	3.2 to 1	1.8 to 1	4.5 to 1
Wakefield Corp.	0.12 to 1	0.75 to 1	0 to 1	0.53 to 1	39.7 to 1	3.1 to 1	[2]	3.5 to 1

1. Excludes financial subsidiaries.
2. Interest expense immaterial.
SOURCE: *Hearings on Conglomerate Corporations*, Part 3, p.158; prepared by the House Antitrust Subcommittee staff.

The ratio of long-term debt to net worth before and after acquisition of eleven major subsidiaries shows that the debt of four of
the companies decreased after merger. The debt of three of those
four, Airport Parking Corp., Levitt & Sons, and Sheraton Corp.,
however, varies according to real estate mortgage requirements.
The other companies show substantial increases in debt after ITT's
assumption of control.

The columns on interest coverage show that after acquisition,
ITT subsidiaries' debts rose more sharply than their revenues. Three
companies—Barton Instrument Corp., Jasper-Blackburn Co., and
Pennsylvania Glass Sand—had more than ten times the resources
for interest payment before merger than after.

ITT World Headquarters, Geneen asserted, increases the debt
of acquired companies after merger in order to provide capital for
the acquisitor's aggressive growth.[5] Those companies' financial
strength declines, then, after merger (see Table 5).

What does the World Headquarters contribute to its 120 acquired companies? Management ability, it replies. But as seen, six of

ITT's nine major acquisitions have experienced decreasing returns from assets after takeover.

Nevertheless, President Geneen told shareholders in the 1971 Annual Report: "There is no question in my mind that the reason for this continued ability to perform under adverse as well as favorable conditions is primarily the result of our worldwide management strength."

We Do Not Manage

The Annual Report averred that management proficiency is the purpose of World Headquarters' control of the subsidiaries:

> From 1959 through 1971 we have steadily increased sales, net income and earnings per share because we have developed a management process which has given our Company a capability we believe to be unique in international industry...More than two hundred days a year are devoted to management meetings at various organizational levels throughout the world... ITT long ago rejected the "ivory tower" approach to management...The test of how well any management performed is simply results. Throughout this wide range of products and services, the common distinguishing element is management...

ITT stated just the opposite to the House Antitrust Subcommittee, i. e., that ITT World Headquarters does not even attempt to manage its acquired subsidiaries. "We do not manage any of the line operations from the headquarters in New York City," Geneen stated on November 20, 1969.[6]

"I wonder whether the Good Lord has given anybody the prowess and the expertise, the ingenuity," Chairman Celler inquired, "to be able to control all those operations in these various fields [under

ITT control] that cover almost every conceivable product, and we are a little concerned about that, naturally, sir."[7]

"I couldn't begin to run one hundredth of what we have, and I don't," Geneen replied. "But I see that it is run."[8]

ITT subsidiaries function as independent "profit centers," he explained. The purpose of the one thousand officials in ITT World Headquarters in New York is to "aid and support" the acquired companies. That headquarters, which had no role in the companies' functions and usually no experience in their fields, somehow gives direction to the approximately two thousand executives who created and very successfully operated those companies before merger. In Geneen's words:

> Each of these people [executives of acquired companies] lay out their own operations, layout their own plans. We review them with them. We insist that they deal with facts and be not over optimistic or, on the other hand unduly pessimistic. Their own plans are what they run against, their own performance is what we monitor with them and then we have this large central staff of about a thousand people which I have mentioned which are there to aid and support them when they get into any problems where they can't meet their own predictions. These people are essentially running their own companies with the support, the help, and monitoring, if you want to call it that, of the central staff, which is a large staff.[9]

Somehow the parent corporation exercises better discipline over the two thousand executives than the discipline they imposed on themselves as independent entrepreneurs. "Let me see if I can make clear what the control is," Geneen continued:

We will insist that they [acquired executives] get the facts and we will insist that they make sensible decisions based on the facts. I am in no position to countermand facts, no matter what my authority is, so they run it. If we had somebody that was not running it well or was proceeding against the facts, let's say his market was disappearing and he was proceeding to build new plants, this is a combination of facts that doesn't make sense and I think we present these facts to him and ask him to make a proper decision.[10]

Executives who founded and operated the formerly independent subsidiaries need the nonmanaging New York office supervisors to tell them to "get the facts." How the acquired profit centers were able to come into being, compete, and prosper without ITT World Headquarters' insistence on "sensible decisions" is another mystery.

Select None But the Finest

"We can hold them [the subsidiary executives] responsible for performance," the president explained.[11] That is to say, successful operations before merger—when the executives had no parental "help, support and monitoring" to fall back upon—required a lesser degree of responsibility.

ITT acquires only prosperous enterprises. To qualify for ITT acquisition consideration, a company must have a pre-merger growth-rate capability of ten percent annually.

The acquisitor would not share Thomas Jefferson's preference for small government. "We are always trying to look for a growing company because we can't manage them ourselves," Geneen explained.[12] Stockholders and public consumers assume the bill for a second, nonmanaging government superimposed over the original, and in all cases, successful, acquired corporate government.

With the invention of the "monitoring" concept, perhaps ITT has added a new horizon to corporate functioning. But credit for the corporation's growth, claimed by the superimposed parent headquarters that neither manages nor directs nor orders may be due to other factors.

First of all, ITT subsidiaries were prospering on their own feet before acquisition. Secondly, the acquisitor feeds from its own mass.

Friends of Friends' Friends

ITT acquired Avis Rent a Car in early 1965. Robert Townsend, president of Avis at that time, suggested to Geneen that not only the ITT syndicate and its thirty thousand employees, but also ITT suppliers (appreciative of ITT patronage) and their employees, serve as assured—if not captive—Avis customers. "We must set up a basis by which we can call IT&T," Townsend wrote, "when we are ready to go after the car leasing business of their divisions, suppliers, friends, etc. and we will get real help...when we are ready to put charge cards into the hands of employees, suppliers' employees and friends' employees we can call and will get real help."[13]

The Avis president's idea fell on fertile ground. Geneen wanted to carry the idea even further. Avis should rely not merely on the consumer power of the thirty thousand ITT employees, he wrote, but also on their power to win friends and influence people. All ITT employees should be ITT salesmen:

> In addition to working out a basis where you can proceed through ITT's people to develop business, it would be desirable to develop a simple basis whereby all of our

employees, to the extent useful, could become salesmen for Avis...We have over thirty thousand employees in the country and if these could be brought in to use it should help. We are awaiting a program from you as to how and in what manner to switch all of our own use over to Avis.[14]

According to Geneen's and Townsend's plan, a car rental agency that is not a subsidiary of an acquisitor with assets of $4 billion dollars and employing the third greatest number of employees in the United States would just have thirty thousand fewer possible customers, and thirty thousand fewer possible salesmen serving without pay. Nor could an independent rental agency count on the suppliers of over 120 subsidiary companies, or those "suppliers' employees and friends' employees" as likely customers.

If a seller of goods or services can make sales from the needs of its employees and their friends and relatives and of its suppliers' employees, of course the competitor that employs the largest mass of consumers and purchases the largest amount from suppliers will be favored. The favored will survive "in self-contained economic domains" (see p. 7), in symbiotic cocoons.

Other companies benefit from that resource, Geneen said:

I don't see anything wrong with our employees believing in our company and being salesmen for it. I think the Telephone Company has people who speak well of them and try to sell telephones.[15]

He meant that American Telephone & Telegraph Co., too, before being dismantled, used its employees as unpaid salesmen.

The scheme of "going after the car leasing business of ITT suppliers" is another example (as seen in Chase Manhattan Bank's outline for Gulf & Western takeovers, p. 27) of an acquisitor's effort to gain reciprocal sales advantages from vastness of operations. Through reciprocity a company forces a supplier to be its customer, not through underbidding or outperforming competitors, but by threatening to curtail its purchases from that supplier. Sales secured by reciprocity, therefore, are not a reward for a manufacturer's quality performance in a competitive system, but rather a reward for the magnitude of its purchases.

But in prepared testimony for congressional hearings, the ITT president expressed his abhorrence of reciprocity in no uncertain terms:

> Efforts to purchase or sell goods or services on the basis of so-called "reciprocity" violate basic management principles of building a business soundly and permanently by being at all times capable of meeting open competition and having a trained business organization capable of meeting the challenges and problems created by competition…The theory of "reciprocity" is therefore repugnant as a basic business philosophy.[16]

Geneen read the "formal ITT Company Policy" against reciprocity, a directive first disseminated to managers in 1966 to make sure that the long-standing formal abhorrence was "clearly understood throughout the system":

RECIPROCITY

> The United States Supreme Court has held that reciprocal buying practices or "reciprocity," is "one of the congeries of anticompetitive practices" at which the antitrust laws of

the United States are aimed. Reciprocity is also an unsound business practice, since it distorts the market process and the normal development of economic efficiencies and product improvements. Consequently, for both legal and business reasons, it is the policy of ITT to purchase and sell products and services on the basis of the commercial criteria of superior quality, suitability, efficiency, service, and price.

No attempt shall be made to develop sales of any service or product through the use of, or threatened withdrawal of, any existing or potential reciprocal buying leverage or "reciprocity."

In this connection, information concerning ITT System purchases from particular suppliers shall not be made available to personnel who are concerned with developing sales and marketing.[17]

ITT does not try to make customers out of suppliers, the policy declares.

Policy Forgotten

But subsidiary managers who forget that policy directive and lapse into reciprocal thinking, as did the Avis executive, apparently need not fear discharge. President Geneen himself, after reading that formal policy at congressional hearings on November 20, 1969, forgot the policy of 1966 the next day. On November 21, at the same hearings, he stated: "Yes, reciprocity effect is an awfully difficult one to define, and I gather a very new term. As a matter of fact, I have never heard of it before the last few months."[18]

The policy statement, however, is not without a most valuable purpose. In 1965 the government sued to prevent the merger of R. J. Reynolds Tobacco Co. with Penick & Ford, a producer of starch used in the manufacture of paper. The government contended that Reynolds could secure sales for Penick & Ford from the paper manufacturers from whom Reynolds purchased in large quantities. In 1969, for the same reason—to prevent market restructuring fertile for reciprocal sales—the government sought to prevent the acquisition of B. F. Goodrich Co., a manufacturer of chemicals and other industrial products besides tires, by Northwest Industries. Evidence was introduced that Northwest Industries and Goodrich's customers and suppliers traditionally traded on the basis of reciprocity. The government claimed the merger would certainly serve to strengthen that tradition. The U. S. district courts denied both government requests for preliminary injunctions against the mergers because both aspiring acquisitors, Reynolds and Northwest Industries, had expressed abhorrence of reciprocal trading in formal, written policy statements.[19] Those statements, the courts decided, overrode abundant evidence of reciprocity schemes.

Similarly, the U. S. District Court for Connecticut, also in 1969, denied the government's request for a preliminary injunction against ITT's acquisition of Hartford Fire Insurance Co. and Grinnell Corp., a manufacturer of water sprinklers. The very same self-serving antireciprocity statement that Geneen read to the House Antitrust Committee and forgot the next day, the court decided, out-weighed the evidence of ITT's practice of reciprocity.[20]

Designs to secure reciprocal sales, as well as other methods for reaping sales advantages from sheer mass of acquired operations—as distinguished from ability to compete—appear in random samples of ITT internal correspondence alone.

The manager of the Wakefield, Ltd., subsidiary, a manufacturer of lighting fixtures, asked ITT World Headquarters in April 1966 to use its influence to secure Wakefield sales:

> A sample special fixture has been submitted to
> Manufacturer's Life Insurance Company, Toronto…Any
> ITT connections that could influence this purchase should
> be employed.[21]

He did not bother to specify the connections envisioned.

A major consideration for takeover of Hamilton Mutual Fund
was that the fund would "tie-in" with and provide captive business for the previously acquired Aetna Finance Company. The
Hamilton Fund, while independent, had developed the business
of loaning over $10 million annually to its shareholders, taking
fund certificates as security. Certain banks in Denver provided
the cash. ITT acquisitions strategists reasoned that once acquired,
the Hamilton Fund could take that business away from the local
Denver banks and give it to ITT's Aetna Finance Co. They wrote
Geneen in May 1965:

> Tony Tyrone of Hamilton Fund was in St. Louis recently, and
> in discussing the possibilities of cooperation between our
> two companies, it was discovered that many of Hamilton
> Fund customers borrow money from Denver banks, pledg-
> ing their Hamilton Fund certificates as security. I suggested
> to him that possibly arrangements could be made whereby
> Aetna could make those loans.[22]

Again Geneen was enthusiastic. Although this was before the
time (according to his testimony) he had heard the word "reciprocity," evidently he was familiar enough with "tie-ins." In a document
actually entitled "Aetna Tie-In," Geneen urged ITT directors to approve the purchase of Hamilton Mutual Fund for that very "perfect
tie-in" reason expounded by the strategists:

Hamilton has increased the practice of loaning shareholders on the value of their fund holdings up to 60% of the equity. This is being done through local banks…It is our estimate that the amount would substantially exceed $10 million per year and is growing annually. This would present a perfect tie-in with Aetna's 170 loan offices for loans on a very secure basis and as a basis of integrating loan revenues and profits from this source, particularly as the fund and the insurance in force doubles and triples in size.[23]

"I didn't write the document," Geneen told the congressmen. "I didn't write it, Mr. Chairman."[24]

Threat to Take Business Elsewhere

ITT Rayonier produces wood pulp used for newspapers. In late 1968 it sought to procure sales reciprocal to ITT Continental Baking Co.'s purchase of Argentine beef. A single Argentine corporation, Transmundo, Inc., like ITT, was both vendor and purchaser. The South American syndicate bought the wood that ITT sold and sold the beef that ITT bought. Rayonier advised ITT World Headquarters on November 19, 1968:

A meeting with ITT Continental Baking Company has been scheduled during November to explore the possibilities of selling more paper pulp in Argentina, using leverage of Continental's large purchases of beef in Argentina.[25]

The term "leverage" meant a threat to curtail Continental's annual purchases of $5 million worth of Argentine beef from Transmundo, Inc., unless its newspaper, Transmundo, reciprocated by purchasing ITT Rayonier pulp wood.

The meeting was dutifully held. ITT Continental informed Geneen on December 23, 1968:

> A group from ITT headquarters and ITT Rayonier met with us on November 22 to discuss potential barter arrangements in Argentina. We purchase over five million dollars worth of Argentine beef in a year. Rayonier is attempting to sell product in Argentina.[26]

A representative of the baking subsidiary met with the president of the beef supplier the following January. According to ITT Continental Baking Co.'s monthly report to Geneen of January 23, 1969, they discussed how Transmundo "could be helpful to ITT in Argentina":

> George Vail met on January 17 with George Senosian, President of Transmundo, Inc. through whom we purchase our cooked Argentinean beef. They discussed in general terms ways in which Transmundo could be helpful to ITT in Argentina. This matter will require further discussion with Transmundo before we can expect tangible assistance.[27]

Geneen contended during the House Antitrust Subcommittee hearings on November 21, 1969, that the discussions with Transmundo's president, George Senosian, were only an attempt to arrange barter transactions to avoid the use of inflated Argentinean currency. He didn't elaborate:

> Mr. HARKINS [subcommittee counsel]. My question is: What did George Senosian do to help them [ITT]?

Mr. GENEEN. I don't know. I don't know George Senosian and I don't know what he did.[28]

Attempts to turn the magnitude of operations to one's own advantage come in various sizes. Some might involve a single piece of real estate in Paris—a "particular little problem which is very small," Geneen said. Other plans might involve capturing multifarious markets upon which hinges a ninety million dollar acquisition.

In 1967 Avis Rent a Car purchased more than $28 million worth of automobiles from Chrysler Corporation. Chrysler was ITT's largest supplier. In September 1968 Avis managers wrote ITT World Headquarters:

> Thanks to pressure on Chrysler we got the garage Dupleix in Paris. This garage is the most sensible proposition which we have encountered in more than two years of searching. It solves our critical fleet service problem for Paris.[29]

Geneen contended that "pressure" was an unfortunate choice of words. But whatever it was that ITT put on Chrysler, it worked. ITT, as the manager said, got the garage.

Levitt & Sons plans, develops, builds, and sells entire residential cities in the United States and abroad. Construction progresses in various geographical areas simultaneously. Levittowns in New York, New Jersey, and Pennsylvania are but the builder's better-known cities. Sales of $72 million in 1966 made Levitt the largest private builder in the world.

In March of that year, the brokerage firm Lazard Freres described to ITT acquisition managers the captive market the acquisitor would acquire for myriad other products upon purchase of Levitt & Sons.

Families could just as well spend 25% more when buying their Levitt houses, the broker wrote.[30] Through such an increase the acquirer would rope them in as purchasers of other ITT products.

That "ready-made virgin market for a complete range of consumer goods and services" would not just be for household products, such as furniture and electrical appliances. Rather, the captive market could encompass the whole range of needs of American family house dwellers—even "all lines of insurance, mutual funds, mortgage services, etc."[31]

The assured customers would not just stop and go. "Levitt creates not only new homes but new households that represent continuing purchasing power," Lazard Freres told Geneen.

Bills Unlimited

Regardless of what the Levitt customer first set out to buy, he would end up with much more than a house. Regardless of which insurance salesman he had been patronizing, he would get the bill from the acquisitor only for "every single item to make a livable home." "The list of items is limitless," Lazard Freres went on, "when one thinks of lawn sprinklers, swimming pools, radio and television, hi-fi equipment...etc."[32]

Better yet, profit from family needs would not only exceed, but might well double, profit from sale of the Levitt-built house. "The gross mark-up and profit margin on this type of business are so substantially higher than those for houses," the brokers' memorandum said, "the total net profit per unit could easily be double that of the house alone."[33]

Even nondiversified Levitt was already benefiting from some of those easy markets. Think, then, the broker implored, what a diversified acquisitor operating in multifarious fields of homeowner consumption could do with such an array of markets:

These sales will give the company a good retail profit margin as the sales facilities already exist at each site. The potential earnings from this program of selling auxiliary items for the home are major and could add measurably to current profits. The eventual scope of such a program is large.[34]

ITT purchased Levitt & Sons for $93,446,000 in February 1968. The housing developer was the largest of all its acquisitions—soon to be dwarfed. In that same month ITT purchased the Sheraton hotel chain for $293,515,000; in April, Rainier Corp. for $302,000,000; in June, Pennsylvania Glass Sand Corp. for $114,000,000; and in September, Continental Baking Co. for $276,000,000. Total acquisition prices for that year, not including the Levitt purchase, easily surpassed $1 billion (see *Hearings on Conglomerate Corporations*, Part 3, p. 32).

According to the Lazard Freres information upon which ITT acted, the purpose of the Levitt acquisition was to secure trade advantages that were unavailable to competitors who were not members of acquisitors' syndicates. Levittown buyers would be manipulated into buying certain brands of goods and services irrespective of the brands' competitive quality. A competitor may well have no alternative but to seek the same advantage for himself by likewise acquiring captive customers through joining a fraternity of acquired companies. If the advantage works (as ITT enthusiastically claims), the dominant corporations will be members of the happy throng. Again, that prediction of a closed economic system brings Darwin to mind. The strength of the surviving fittest will come not from their ability to provide superior goods and services but from their manipulation of suppliers, suppliers' friends, suppliers' employees, subsidiaries, subsidiaries' friends, subsidiaries' employees, subsidiaries' employees' friends—and the ever-expanding citizenry of Levittowns.

The profit from the Levitt scheme flowed beyond the acquisitor. The benefit to Lazard Freres for locating those captive customers was a $250,000 acquisition finder's fee.

Excluding that amount, between January 1, 1966, and September 5, 1969, Lazard Freres received $2,075,000 for ITT merger services. For other brokerage services, Lazard Freres received an additional $1,792,790 from ITT. In 1968, 34.2% of the firm's acquisition-fee income involved ITT mergers.[35]

In public testimony and with impeccable authority, Mr. Rohatyn, a Lazard Freres partner and director who was also an ITT director and ITT executive committee member, described a system of interlocking directorships that keeps the acquisition process well greased. At least one of the brokerage firm's partners was on the boards of directors of twenty-seven of the companies involved in the sixty-eight mergers the firm arranged from 1964 through September 1969.[36] Rohatyn first conceded and then denied that such interlocking board service helped the firm get the business.[37] But with no equivocation he stated that any failure of Lazard Freres to secure the business of a company on whose board a Lazard Freres official served would be "to our chagrin."[38]

Interlocking directorships enabled Lazard Freres to perform better service, Rohatyn said. "I would certainly hope that if one of our partners is a director of a company, that we would be…selected to perform the advisory service."[39]

Caesar's Wife

Lazard Freres was not disappointed. The firm's income from mergers of companies with which it "locked" at least one directorship increased from $841,030 for the entire year 1964 to

$2,217,802 for the first three quarters only of 1969.[40] The system is unbeatable, works just as well in reverse, and feeds on itself. The firm acquires merger business through interlocking directorships and acquires interlocking directorships through merger business:

> Generally corporate clients sooner or later will invite one of our partners on the board, because really this is the way it happens...If we had dealings with a company and have performed services, by and large, at some point or another we will be invited on the board, and the relationship may become close.[41]

Lazard Freres has also benefited from arbitraging the stocks of the companies involved in the mergers it initiated. To arbitrage the stock of merging companies is to purchase stock that is expected to rise (usually the target's) and sell short stock expected to decline (usually the acquirer's), as a result of the merger. A broker who participates in an acquisition may lawfully arbitrage the stock of the two companies only after the announcement of the merger. Ostensibly, the public had the same opportunity Lazard Freres had to profit from arbitraging stocks of the companies the brokerage firm helped merge.

To sell short is to attempt to profit from the decline of a company's stock, and hence from the deterioration of its performance (see Goldman Sachs, p. 3). Lazard Freres states that it has therefore adopted a rule of self-restraint against arbitrage of mergers involving companies for which a Lazard Freres partner serves as director. The firm never places itself in a position of hoping for a decline in performance of a company with which it interlocks directorships—the firm says.[42]

Nevertheless, partners of Lazard Freres gained $236,321 from arbitraging the Rainier-ITT merger by purchasing Rainier stock and selling ITT short.[43] A Lazard Freres partner, Rohatyn, was a director and an executive of ITT at the same time.

The rule presented no obstacle, Rohatyn explained, because the firm's holding in the two stocks was distributed to all the partners individually, except to ITT director Rohatyn.[44] But the partners thereby benefited the same as if the firm still held the stock. The rule designed to prevent selling short the stock of companies with which Lazard Freres interlocks directorships, obviously, then, can be circumvented at will. But by virtue of the rule, Rohatyn stated, "we have always managed to be absolutely purer than Caesar's wife."[45]

Huge Salary Increases

Another system which, like interlocking directorships, establishes rapport among parties to merger negotiations is the promise by acquiring officials to raise the salaries of the acquired managers. *Forbes,* May 1, 1968 (p. 28), reported:

> Geneen, incidentally, uses money to attract not only people, but companies as well. His favorite technique for making sure that managements will favor being acquired by ITT is offering management five-year contracts and huge increases. His recent arrangement with Sheraton Corp. is typical. ITT has entered into five-year contracts with Sheraton's top three officers. The trio will be paid a total of $315,000 a year vs. their pre-ITT salaries of $172,000. Small wonder so few managements object to being acquired by ITT. Happiest of all are executives about to retire. For they can stay on the ITT payroll at substantial increases,

when they would have had to live on some slender pensions otherwise.

Officials of Continental Baking Co. received salary increases of 50% immediately after merger. Officials of Rayonier received increases of almost 50%. Officials of Avis received increases of 100%. The three highest salaries paid by Sheraton Corp. were increased upon merger as follows: $55,000 to $85,000, $56,000 to $105,000, and $61,000 to $125,000.[46]

But the ITT president disputed that he used salaries to gain companies:

MR. GENEEN. These mergers are voted on by the stock-holders. They make their own individual decisions as to whether they want to come with us or not.

Chairman CELLER [of the House Antitrust Subcommittee]. I can 't disagree with you more. If managements are given substantial increases, they undoubtedly can have their effect upon the objectivity of management's recommendations to the stockholders. There is no doubt about it.[47]

Salaries of acquired company officials were public information prior to the mergers. But Geneen asked that the subcommittee not disclose the list of salaries submitted to it:

My own feeling is that this is personal information to the individuals and companies. Unless it serves some useful purpose, I would request the Committee to keep it confidential.[48]

"Why should this type of information be available [through the SEC] when a company is independent," the subcommittee counsel asked, "but not available to the public after it's acquired?"[49] He received no reply.

CHAPTER 9

The Purpose of LTV, Inc.—Breaking Companies Apart

Not all parent acquirers attempt to use the size of their operations to secure anticompetitive advantages. Some parent headquarters appear to exist only to continue to acquire.

James Ling, the founder of Ling-Temco-Vought, Inc., was CEO during the five years when it acquired over $2.817 billion of other companies' assets. His relationship with management was severed in mid-1970 when the banks questioned the company's ability to repay its acquisition loans and interest. The stockholders changed the corporate name to LTV, Inc.

James Ling's memorandums written during the height of the acquisitor's takeover campaign describe the management purpose for piling up other companies' assets. It elevated LTV from the 285th largest U. S. industrial corporation in 1960 to the twenty-fifth largest in 1969.

A Parent's Gift

Paradoxically, Ling's purpose was to leave the acquired company as if it had never been acquired. LTV instructed its managers that the consolidated companies were to function on their own. Of the services the parent headquarters performed, foremost was "external growth," i.e., acquisition of more companies. "Internal growth," i.e., growth by any means other than acquisition, was strictly the responsibility of subsidiary, not parent, management.

When he began taking over corporations, Ling conceived "operating umbrella versatility." He wrote division executives on June 14, 1964:

LTV Corporate Headquarters Operating Philosophy

A distinct advantage materializes to the benefit of the division in that the division has most of the fringe benefits of a small business and yet retains the operating umbrella versatility and strength of the LTV corporate structure.[1]

Those benefits, Ling told the House Antitrust Subcommittee in April 1970 were that:

LTV furnished corporate services to them [the subsidiaries] in industrial relations, public relations, and so forth. On the other hand, from their point of view, the fringe benefit was that they would have quick reaction capabilities to their customers, as opposed to going through the morass of the big bureaucratic corporate structure which undeniably exists at times. In other words, they had the advantage of being a small business by comparison but on the other hand, had the advantages of a parent company which could give full support to them by providing legal, accounting and planning functions for the benefits of the underlying subsidiaries.[2]

Why the parent company could perform legal, accounting, planning, and public relations functions for the subsidiary more adequately than the subsidiary itself, he did not say.

After acquiring $175 million more of assets, Ling revisited the idea that subsidiary operations were quite naturally the subsidiary's concern. The parent was to be left free to negotiate more acquisitions. On October 16, 1967, he told company officials:

> Within the area of internal growth...the primary responsibility for securing that growth belongs to each of the subsidiary presidents and his management. It is LTV's desire and expectation that *subsidiary management will concentrate its time and efforts* on the development of products, processes, services, marketing and distribution, production and inventory control, and administrative and financial control to ensure the planned internal growth.[3]

That encompassing list of a subsidiary's duties might leave for the acquirer a function of counseling. "Ultimate responsibility" for management goals, however, clearly lay with the subsidiary management, he continued:

> LTV corporate staff will monitor, counsel and assist within the scope of our particular expertise in the settling of operating plans, financing of expansion, solving operating problems and, in general, the attainment of internal growth goals, but the ultimate responsibility for attainment of these goals is that of subsidiary management.

The parent company's real duty was to acquire more corporations. Mr. Ling explained:

> It is, and has been, a stated objective that LTV and its subsidiaries will grow externally to supplement planned

internal growth. Growth in this context, as has been stated previously, is simply defined as an increase in earnings per share for all the shareholders of LTV and its subsidiaries which results in a higher intrinsic, or market, value for the shareholders' security.

Within the area of external growth, the sole responsibility for planning, negotiating, consummating and integrating acquisitions and mergers for LTV and all its subsidiaries belongs to the LTV corporate staff. More specifically, the total responsibility within this area belongs to the office of the Chairman of the Board and Chief Executive Officer. This responsibility has been delegated to the Vice President-Corporate Financial Plans for implementation in all its phases, but its ultimate responsibility is mine.

Forget Loyalty

Two years later, after having acquired $913.8 million more of assets, Ling told the chief executive officers of the subsidiaries that their divisions were to become even more "operationally self-sufficient":

LING-TEMCO-VOUGHT CORPORATE INTEROFFICE CORRESPONDENCE

October 14, 1969

To: See Attached Distribution

From: James J. Ling.

In my closing comments made during the recent planning session at the Ranch, I suggested that you submit any suggestions you might have concerning same. We here at Corporate are convinced that a change in format is in order, particularly in view of the new companies which will be coming on stream shortly. So send your suggestions directly to me—we can discuss as time permits.

Last year and again this year we stressed the goal that all the underlying subsidiaries should become operationally self-sufficient as soon as possible... Oversimplified, the Chief Executive Officers of each of the subsidiaries will have complete responsibility for operations and profit performance—*the buck passing stops at the CEO level of each company!*

I'm looking forward to an "on target" report from you at next year's off-site meeting.

JAMES J. LING[4]

Though the benefits to the acquired companies from takeover might be ambiguous, those companies' duty to their new parent was clear. Loyalty of the acquired managements was owed first to the parent acquirer and only secondarily to the companies they managed. Ling urged managers who had been transferred to the "remote" LTV Tower in Dallas to discard feelings of loyalty to their companies:

June 18, 1964.

To: LTV Corporate Staff Heads and Division Managing Executives

Subject: LTV Corporate Headquarters Operating Philosophy

We have completed a most significant step in establishing our Corporate Headquarters in a separate location and as a compact unit here at the LTV Tower. This move to a central location, remote from any major division, for the Corporate group reflects a business strategy and an operational philosophy...We, at the LTV Corporate level, must more completely eliminate prior feelings of allegiance or emotional stimulus that could make us subjective as a result of our background ties and experience with any particular division. We must become completely unemotionally detached from our heritage in this respect.

The rationale and vital necessity for doing so is quite clear. Our responsibilities are clearly set forth. We must produce a profit result for the Corporate owners which will provide increased earnings per share and will build value per share on a long range as well as a short range basis.[5]

Profit to the parent company, yielding greater earnings per share, yielding higher LTV market value, yielding greater resources for acquiring more companies, was the goal.

Favor Some Divisions Without Favoritism

Four days later, Ling issued a clarifying memorandum. An executive's loyalty to the acquired subsidiary, developed and nourished over the years—usually thought to be an asset—is instead a liability. A subsidiary, he said, is above all a salable commodity. Too much attachment for an acquired company might impede the acquisitor's chances of reselling it for the best price:

The rationale for pushing the point about becoming emotionally detached from the prior heritage of a division is as follows: Charting the course of LTV over the next several years will call for various and vigorous actions in organizing and reorganizing, personnel placement, apportionment of R&D/BP&E funds and favoring emphasis on some divisions more than on others. We can play no favorites.

To illustrate: If it were to our best interest to sell off a division or product line and redeploy the assets in a more rewarding manner, then this we must do.[6]

To the House Antitrust Subcommittee Ling stated, nevertheless, that the sale of a company can be disruptive to its operations. LTV had recently announced plans to sell Okonite Cable Co. for the third time in twelve years. "Apparently after the acquisition," he explained, "there is a deterioration of morale."

Mr. HARKINS [Subcommittee counsel]. Do you think that there has been a disruption of its [Okonite's] operation resulting from LTV's purchase of the company in 1966?

Mr. LING. Well, did we disrupt the company?

Mr. HARKINS. Yes.

Mr. LING. I think from that point of view the record is quite clear there was a possible disruption.[7]

In spite of that admission, the LTV chief executive saw no reason why acquirers should not trade their acquired companies as salable commodities:

Chairman CELLER. Suppose you were a shareholder of
Okonite and it was purchased by Kennecott [acquirer of
Okonite in 1958] and then sold by Kennecott to LTV who
subsequently sells it to somebody else. What would be your
attitude as a shareholder?

Mr. LING. My attitude would be, what price did they pay?
They could not do all those things without buying up securities. If they offered enough money, I would be like the
shareholders have been; I would probably sell it.

Chairman CELLER. If I were a shareholder, I would kick
like hell.[8]

LTV sold Okonite Cable Co. in 1971 to Omega-Alpha
Company, a holding company which James Ling later managed.
Also Braniff Airways was to be sold to comply with a judicial con-
sent decree whereby LTV would be permitted to retain control of
Jones & Laughlin Steel Co. In 1970 and 1971, in order to pay ac-
quisition debts, LTV sold three of its most profitable acquisitions:
Wilson Sporting Goods, Wilson Pharmaceutical Co., and Allied
Radio. Previously, LTV sold its controlling interest in National Car
Rental, Stonewall Insurance Corp., First Western Bank & Trust,
and American Amicable Life Insurance Co. The banking and insurance interests amounted to over $100 million.

LTV internal reports show that an acquisitor that leaves operations entirely to the acquired managements does not fare as well as
one that controls its subsidiaries so as to gain competitive advantages
through the size of the combined operations. A mere comparison of
International Telephone & Telegraph Corp.'s income statement with
LTV's statement, which shows deficits for 1969, 1970, and 1971,
should be indication enough.

The U. S. Army arranged for LTV's acquisition of Memcor so that the acquisitor (according to the Army Contract Adjustment Board decision of September 12, 1966) would correct Memcor's "lack of competent management."[9] But LTV managers' reports leave doubt that the parent headquarters that told its subsidiary officers, *"the buck passing stops at the CEO* [subsidiary chief executive officer] *level of each company!"* attained that goal for its Memcor subsidiary. An early February 1967 level managers' report to LTV headquarters states that Memcor profit was $8,000 above "target" only because of the army's gratuitous payment increases, and that sales were $6.3 million below target. A few weeks later, the reports continue, 30% of Memcor's PRC-25 production was rejected.

LTV informed its shareholders in its divisional Electrosystems 1967 Annual Report that the Memcor merger had strengthened the acquisitor "management and scientific staff" and "high-level technical skills." However, internal reports that same year show that LTV itself was experiencing exactly the same problems that the army expected it to correct in its acquired company. According to managers' reports later in 1967, manpower problems even under the acquisitor's management prevented Memcor from producing its backlog of orders.

The internal report of March 25 explained that the major manufacturing problem—utilization of available manpower—had become "grim."

The report of April 24 stated that the problem had not abated and that Memcor, as before merger, could not produce the business it had contracted.

On May 30 the managers reported no improvement: "23 failures to date and all of them catastrophic."

On June 25 the Memcor managers sent word to the LTV Tower that electronic equipment being produced for the Italian Navy did not meet "specification." They urged that LTV try to persuade the Italians to sell the failing equipment to a NATO command.

The acquisitor's vaunted ability to aid distressed subsidiaries was to no avail, or did not exist. On August 28 the managers related that although LTV president Clyde Skeen had advised Memcor to draw on the assistance of the other LTV subsidiaries, its deficiency in engineering talent persisted.

The managers' report of October 24 stated that not even available manpower was being utilized. Referring to "serious degradation of the application of senior engineering talent," it concluded: "The loss will grow, possibly double."

Unlike those confidential reports, LTV public earnings statements after the early 1967 Memcor merger revealed no sign of the management problems. The acquisitor's Annual Report of 1967 stated that the acquiring division's [Electrosystems'] profits were up 102%, from $2.6 million in 1966 to $5.3 million in 1967. LTV's calculation of the good bookkeeping effects derived from the pooling of revenues of the merging companies had come true. A footnote added, however, that Memcor profits had been added to the division's 1967 (post-merger) profit total but not to the 1966 (premerger) total. Thus, in announcing a 102% profit increase, LTV was comparing the combined profits of two companies with the profits of one company. In 1968, when combined profits had to be compared to combined profits of the previous year, the difference became a 32% decrease as opposed to the 102% increase in 1967.

By adding the income figures of its acquired companies to its total income figures, an acquiring company reports rising income. By reporting rising income, it heightens the market value of its stock and, in turn, obtains greater power to continue acquiring other companies.

In December 1967 LTV purchased General Felt Industries, Inc., a manufacturer of floor covering. The acquisition's revenue and earnings were added to the figures of the previously acquired Okonite Cable Co. During the first year of LTV control, General Felt's sales rose dramatically. So much, Okonite's $8 million decline in sales in 1968 was reported to the public after combination with General Felt sales as an increase of 85%. LTV reported to Okonite shareholders in the 1968 annual report:

> The year 1968 was a mixed one for the Okonite Company. While certain of the year-end figures proved disappointing, there was progress. Much was accomplished. Much was begun...Sales rose to $190,848,000, an increase of 85% over the prior year's mark of $103,238,000.

The disappointing figures were that Okonite's pretax earnings had decreased from $16.4 million in 1967 to $12.1 million in 1968 (even with the addition of General Felt's earnings). Again, the management's internal reports are not consistent with its public reports. In the 1968 Annual Report to shareholders, LTV said that the reasons for the profit decline were "lack of expansion in utilities and industries" and the copper strike during the first part of the year— reasons absolving the parent of blame. Any public acknowledgment of responsibility for the profit decrease, and thus of management deficiency, would likely have caused a decline in the market value of LTV and Okonite stock.

It Does Not Make Sense

The managers' internal reports, however, explicitly refute those public explanations. They cast doubt on ability to compete:

> 24 June 1968: During the recent meeting, it was stated that the cable industry was way off. This seems reasonable in the

case of construction, but for manufacturing, OEM and the utilities, it does not make sense.

...[N]ow that the copper strike is over and producers' copper is now available, it might have been expected that the Okonite situation would have improved substantially. However, such is not the case; rather their situation has worsened.

A managers' report of July 10, 1968, cited "severe competition" as the reason for sharply declining Okonite sales and profits. Though amalgamation of income figures may result in higher earnings on paper, amalgamation of operations of industrially unrelated companies (the cable manufacturer and the floor covering manufacturer) does not result in added financial resources, the July managers' report indicated:

> There is a higher level of operating cash requirements for General Felt as compared to what Okonite has been used to. As a result of their cash shortage, they have added $7 million short-term credit. Okonite has stopped commitments for any capital improvements or maintenance until the cash situation improves. The biggest factor concerning shortage of capital is the fact that the Santa Maria facility [an Okonite plant] has not been sold and leased back.

Split Ownership
LTV typically sells off minority interest in acquired subsidiaries. LTV retains the majority or controlling interest (more than 50 percent of the voting stock) of several subsidiary companies while their minority interest is traded publicly on the American Stock Exchange. LTV has divided Wilson Meat Packing Co., which had

been traded as one entity on the market before acquisition, into seven separate publicly traded companies. By that system, which Ling denominated "Project Redeployment," the acquisitor obtains control of the subsidiary without having to expend capital for more than 51% of its voting stock.

That division of subsidiary ownership by a parent company exercising control of the publicly traded subsidiary raises issues of conflicting interest. For example, Wilson owners who chose not to accept LTV's generous cash tender for their stock suddenly found themselves minority shareholders of a company controlled by LTV. Their preferences and interests conflicted with the interests of the acquirer, which canceled their Wilson common and issued them LTV preferred shares—an exchange those stockholders would not have made voluntarily.

LTV's practice of selling off minority interest in subsidiaries was becoming a growing trend, *Forbes,* January 1, 1972 (p.125), reported:

> Ling's specialty at LTV was not only acquiring companies, but also rearranging them and splitting them off into independent, partly publicly owned enterprises. "Redeployment" he called it in 1965 when he spun off LTV Aerospace, LTV Electrosystems and LTV Ling-Altec, and again in 1967 when he broke Wilson & Co. into three independent parts.
>
> Curiously, few imitated Ling's strategy in LTV's days of glory, but now that L-T-V has collapsed, many more have adopted it…For such companies, the advantages are considerable. In selling 30% of Globe Systems, for instance, Kidde generated some $11 million in much needed cash, converted an asset with a market value of about $30 million.

The creation of a public market has its advantages, notably in easing the disposition of added shares in the future.

Forbes foresaw clashes of interest. A policy designed to prevent conflict between LTV and minority shareholders restricts parent company officials from owning shares of subsidiary companies.

The minutes of the LTV board of directors meeting of May 9, 1968, provide:

RESOLVED, That it shall be the policy of Ling-Temco-Vought, Inc. to prohibit its officers and directors from owning any securities issued by any of its subsidiary corporations, as well as the securities of any other companies in which LTV subsequently may acquire control; provided, however, that this policy shall not prohibit any person who may be both an officer of a subsidiary corporation and an officer or director of Ling-Temco-Vought, Inc. from owning any securities of the subsidiary of which he is an officer through the acquisition of stock by open market purchases or by exercise of stock options.[10]

"This is an old policy," Ling explained to the House Antitrust Subcommittee, "which was put into effect in 1965, in order to avoid conflict of interest."

As an illustration, I do not own any securities in any of these underlying LTV subsidiary companies, because the policy forbids it. It provides I can't do it. On the other hand, Mr. Lawrence [president of Braniff] is a director of LTV, and thus should have an interest in LTV. On the other hand, his principal endeavor is in the future of Braniff, so thus he would be entitled to have an equity position in each company. Thus, oversimplifying, the reason for the policy is to eliminate the possibility of conflict of interest.[11]

People Are Honest

Ironically, LTV and the New York Stock Exchange, while guarding against conflict of interest so well that both disallow the remote power that a parent company official might hold over a subsidiary (of which he is not an official) through his trading of its stock, absolutely allow the parent company the power to name the entire boards of directors of all its publicly traded subsidiaries.

Voting rights of securities of the LTV subsidiaries, with a few exceptions, are noncumulative. LTV, therefore, if it chooses, can name their entire boards of directors, filling them with its own directors or other representatives, by virtue of its ownership of at least 50% of each subsidiary's voting securities.

Such control of the subsidiary boards empowers LTV to call its callable loans to subsidiaries at any time, though to their detriment; to raise money for acquisitions by issuance of equity or debt securities of the subsidiary—and to decide for itself the amount of earnings the publicly owned subsidiary retains or pays to LTV in dividends.

As an example, LTV planned to raise capital for acquiring Wilson & Co. by calling a $10 million loan, not due for two years, owed by Okonite. If interest rates had risen, of course, Okonite could have replaced the capital only at greater expense. Also, LTV planned to issue debt and equity securities of other subsidiaries.

The plan was not adopted. Okonite did repay the loan fourteen months before maturity of the note but replaced the funds by borrowing at the lower rate of 4.75%. Ling acknowledged that LTV's power to call subsidiary notes, or any other power over subsidiaries resulting from the parent's authority to place whomever it chose on the subsidiary boards, could possibly result in conflict of interest "because nothing is impossible." But, he continued, "all subsidiary directors...have to act in behalf of the minority shareholders as well as the management" whether they are chosen by the majority or minority stockholders, because "whatever is good for the management is good for the minority, and vice versa."

He explained:

> There could not be a problem because of the full visibility of these companies. Being publicly owned, and transactions are duly reported between the companies, and all transactions are reported in various and sundry agreements. Even if one were to have the temptation, one would resist it, because again, I think, most people, being basically honest, will make the proper decisions for the betterment of that particular company.[12]

If LTV were to act against the interest of a subsidiary, such as to call its loan and force the subsidiary to reborrow at higher interest:

> ...we would be liable to the minority shareholders and the independent members of the companies whose incentives for the future depend on that company. On the other hand, if you are saying that LTV can be irresponsible, well, anybody can be irresponsible. Conflict of interest among directors who are on both parent and subsidiary boards really hasn't developed.[13]

Ling eventually acknowledged, however, that LTV had appropriated to its own use proceeds from the sale of subsidiary securities, and that LTV had converted subsidiary shares so that those securities began to pay dividends to LTV. He explained:

> You see, we have never sold any stocks… that I can recall, where LTV itself got the proceeds, you see. We have never done that. The underlying companies, in every instance, have received the funds for their own particular business purposes. We have never sold any Wilson stock in our behalf, or any Braniff stock, *until recently*, in our behalf, and so forth [emphasis added].[14]

That LTV converted Class B nondividend-paying stock of a majority of its subsidiaries into dividend-paying securities "is correct," the LTV chief executive stated. The earnings represented by that stock thus flow to the parent company. Ling continued:

> We did convert. In the original stages of Project Redeployment, we wanted the underlying companies to have the advantage of keeping all the funds generated through earnings for their benefit, so only the public shareholders received dividends with our objective through the years, as the subsidiaries grew, being that we in turn would participate. We did not participate for some period of years. This past year [1969], we have started participating through exercise of conversion provisions of a certain amount of the securities.[15]

In short, whether the publicly traded subsidiaries retain their earnings for their own development or pay them to their majority shareholder is wholly determined by that shareholder.

Dark Visibility

The conversion of formerly independent corporations into subsidiaries of an acquisitor obscures "visibility." An independent, publicly traded corporation must adhere to the Securities and Exchange Commission's disclosure requirements: it must expose to public investors' view information about its profitability and stability. That information shows quality of management and enables the public to make more "informed investment decisions." Most acquisitors do not reveal the income and stability figures of their acquired corporations after merger. The annual report of ITT, for example, does not show the revenue and profit figures of any of its more than 160 individual acquired companies. But each of those publicly traded companies publicly reported those figures before merger. The ITT report discloses revenue and profit figures only for the nine general divisions into which the companies have been amalgamated. Thus, standard investment information accessible to the public before acquisition is not available afterward.

LTV, however, contends that its "redeployment" affords the public better insight into operating conditions than other acquisitors afford. Because the "redeployed" companies are publicly sold, they each must satisfy SEC disclosure requirements. LTV asserts that the public has a better view after acquisitions than before. The 1967 LTV Annual Report told shareholders:

> The term "'visibility'" indicates openness to public view. In a business sense, visibility leads to easy recognition of outstanding or sub-par performances. In this regard, we believe LTV is one of the most visible companies in the United States today. None of this is to say that we in any sense consider ourselves infallible. The perfect corporate structure does not exist, and it is likely that it never will...

Recognizing this, we have attempted not only to diversify LTV, but to make its component parts clearly visible.

To the House Antitrust Subcommittee on April 15, 1970, Ling elaborated: "Thus, it is not really too surprising that *Fortune* magazine declared in June of 1968 that LTV was the most visible American company out of the 500 or 600 listed in their Fortune Review."[16]

That reasoning itself is obscure. As already seen, the public was led to believe that the publicly traded Okonite subsidiary had increased its sales by 85% in 1968, while that increase came entirely from the simple addition of General Felt sales to the Okonite total. Reporting of income was no less obscure. The LTV and Okonite annual reports to shareholders showed only a decline of pretax Okonite profits from $16.4 million in 1967 to $12.1 million in 1968. Nowhere did they show that Okonite pretax profits, without the addition of newly acquired General Felt profits, declined $11 million, or 73%.

Textbook Shredded

Eventually, the acquisitor's total earnings failed to increase even with the addition of acquired earnings. LTV's net income in 1965—the last year before 1969 in which it made no acquisitions—amounted to $6 million. In 1967 income rose to $34 million. But although the incomes of the four companies acquired in late 1967 and 1968 were added to the 1968 figure, income fell to $28 million. In 1969 it fell to a $38 million loss (before estimated future income tax benefits).

Obviously, the acquisition program did not increase return on assets. In 1964, total revenue equaled value of assets multiplied by 2.55. In 1969 that multiple had decreased to 1.28. In 1964, net income equaled 3.9% of assets; in 1968, 1%; and in 1969, of course, a negative figure.

In April 1970 LTV's income was not sufficient to pay the interest on its acquisition debt. James Ling was asked whether that debt, soon due, exceeded LTV assets by over $150 million. The crowded hearing room was silent. Either the audience would hear the master of take-over finance admit that at last his back was to the wall, or hear another explanation of corporate restructuring that shatters textbook rules:

Mr. HARKINS. Does not in fact the debt of LTV exceed its salable assets by $170 million?

Mr. LING. I am sure you will want a very honest comparison. If I may go to the chart.

The CHAIRMAN. What was your answer about the parent company? Is it true that LTV's parent company long-term debt exceeds its salable assets?

Mr. LING. No. If he is going to evaluate the stock market on the asset side, he has to stay with the market on the debt side. Our calculations show we have $210 million assets over debt. The debt traded on the market does not trade at $800 million some odd but trades on the order of $400 million some odd.

Mr. HARKINS. But the debt is a legal obligation as of the time carried on your books.

Mr. GRIFFIN. That is correct.

Mr. LING. We have to pay it eventually but we are not obligated to pay it today, by any stretch of the imagination. On the other hand, we can buy it at 42 cents on the dollar as of yesterday.[17]

Each dollar represented by LTV debt securities was selling on the market at 42 cents because investors had little confidence in LTV's ability to repay the debt. The acquisitor itself, then, could purchase its debt securities, as Ling said, for far less than it had convinced the public to pay for them.

By that strategy a corporate acquirer first borrows the money from the public to purchase other companies. Supported by those acquisitions, it taxes but does not contribute to their operations. The acquired companies are less profitable then under the acquisitor's control. Public holders of the debt securities it sold in order to purchase the companies lose confidence in its ability to repay the debt. They thus sell the debt securities for 42 cents on the dollar. The acquisitor, if it has the resources to buy them back even at that price, then pays off its debts for less than half the amount it borrowed. Hence, the acquisitor profits, and augments the trend of corporate concentration, by its management failures.

"I will capitalize from my inability to repay my loans," Ling, in effect, said. The government apparently agreed that the LTV chief had hit upon a smart scheme. The Treasury Department declared such profit from repurchase of corporate securities below their sales price to be nontaxable, on January 20, 1977. "Thank God for Treasury," one tax counsel said (*The Wall Street Journal*, January 22, 1977).

Goldman Sachs should know, then, it has government sanction for its shorting of securities it packages, sells, and buys back below sales price (see p. 5). The Treasury Department has not ruled, however, that Goldman Sachs' profit from the buyback is likewise nontaxable.

CHAPTER 10

The Purpose of Litton Industries, Inc. —Winning Government Money

We don't want anybody going up and down the highways and byways blasting out this information. This is the reason we are in the spot we are in here this morning. We are trying to get facts, not headlines. That is why we have the doors locked to the news media. I hope the mouths of the members of the committee will be closed to the news media.

> Chairman F. Edward Hebert at outset of the *House Armed Services Committee Hearings on Delays and Cost Overruns of Ingalls Shipbuilding Corp.*, Armed Services Hearings, p.10575.

Litton Industries says its management ability is transferable to the multifarious industries it has entered by acquiring other companies. Its managers say they can operate those companies without training in their industries. That the acquiring management is capable of operating myriads of industries without experience before merger is the most common of acquisitors' claims. It is Litton's explanation for having acquired 119 companies in the nine years between 1960 and 1970. During that time Litton's total assets increased 1,228 percent, from $119 million to $1,580 million.

Litton's management even improves the performance of acquired companies operating in industries outside of its experience, it says. Yet the acquired managements are indispensable to the acquisitor. Glenn McDaniel, chairman of the Litton executive committee and former president of Radio Corporation of America, explained to the House Judiciary Committee on March 4, 1970:

> In our acquisitions we value good managers. We look to the existing managers for their specialized expertise and supplement them with Litton's broad management capabilities. We can develop their enthusiasm and expand their horizon...
>
> Corporate central management works with each of our individual managers in periodically setting business goals and the main strategies for realizing them... Like a democracy, our organization is not perfect in its operations...[1]

Litton's management of acquired subsidiaries strengthens the competitive process, Mr. McDaniel said:

> We attempt with careful strategic planning to improve and broaden the products of acquired companies through the use of advanced technology.
>
> This is a long term process. We cannot emphasize too strongly that the process of strengthening acquired companies in this manner takes many years of product development and intense managerial effort...
>
> We have introduced beneficial competition into a number of industries.

Litton's performance of defense contracts tests its claim to transfer management techniques to untried endeavors. Military and naval production makes up most of Litton's operations, as it does operations of other serial acquisitors. Litton's Defense and Marine Group derives 97% of its revenue from government contracts. That group's revenue, in turn, amounted to at least one one-fourth or one one-third of Litton's total revenue during the period 1968 to 1972.[2]

Big Ship—Small Competition

Whether defense production standards influence Litton's entire mode of operations is the question. The House Judiciary Committee asked Litton Chief Executive Tex Thornton that question four times:

> Has it been Litton's experience in dealing with the Government that delay, confusion, or resistance to requests tends to work to the advantage of the contractor?
>
> Does Litton employ techniques and procedures in performing Government contracts different from those required for doing business with private industry?
>
> Does the mere passage of time work to an advantage of the contractor in dealing with the Government?
>
> Has it been Litton's experience that turn-over in Government personnel administering Government contracts and the resulting delay is advantageous to Litton?[3]

Fortune, April 1968 (p. 139), had raised much the same questions when it reported that the government accepts a less competitive, lower standard of performance than do defense contractors' commercial clients. It said defense contractors profit from malperformance:

> The requirements for profitability in government work are less exacting than those of the private marketplace. In

the advanced government projects where Litton has made its special mark, success depends almost solely on performance. Barring flagrant mismanagement, the company that can do the job can be reasonably sure of clearing a respectable profit. Minor delays and mistakes of judgment can be overlooked, and contracts are drawn to allow for unforeseen snags in research and development. Private customers are less forgiving, largely because in most cases there are competing suppliers of roughly similar products.

Pertinent to the committee's four questions—and to the claim of adaptability of Litton's management ability to untried endeavors—is the acquisitor's operation of Ingalls Shipbuilding Corp. of Pascagoula, Mississippi. Litton had never built ships before it acquired Ingalls. Yet Litton expressed a primary reason for the purchase: opportunity to apply its electrical technology.

When Ingalls was first offered for sale in 1961, *Time,* October 4, 1963 (p. 107), reported that Litton's chief executive, Tex Thornton,

> brooded over the possibilities for weeks, finally concluded that the nuclear submarines that Ingalls was building were really just a collection of electronic machines and devices packed into a hull, and therefore an excellent destination for the products of Litton's expanding electronics complex.

Six years after the $20 million purchase, shipyard manager Ellis B. Gardner informed Roy Ash, Litton president, of Ingalls' inability to secure profits through the rigors of competitive bidding, of miscalculation of the true magnitude of operating losses, and of "the degree to which the…facilities have been allowed to deteriorate." He wrote on June 7, 1967:

> The operating loss experienced by Ingalls during 1967 and our obvious inability to have been able to predict its true magnitude I realize can produce a most skeptical frame of mind…
>
> The situation in which we now find ourselves does not lend itself to successful strategic bidding for maintaining a continuing influx of reasonably profitable new business…
>
> Another serious limitation at Ingalls is the degree to which most of the general purpose facilities have been allowed to deteriorate. Piers and shipways require major overhaul. We have just completed a comprehensive review of our electrical distribution system, the results of which indicate that much of this equipment will soon become completely undependable unless replaced, due to age and lack of care over the years. Of the ten original shipways at Ingalls, three have been unusable for a number of years and a fourth needs about $900,000 of repair work…[4]

An alternative, however, to incurring expense by merely replacing Ingalls' limited facilities, Mr. Gardner continued, would be the more daring project of constructing a second, entirely new and larger facility for the construction of larger ships on the west bank of the Pascagoula River, across from the original yard. Competition for the building of smaller naval vessels was very intense, the memorandum read. But the larger the vessel, the fewer contractors. He wrote, "For certain classes of large complex ships such as aircraft carriers [Newport News] no longer has any competition."[5]

Ingalls presented no challenge to major shipbuilders such as Newport News and General Dynamics, the manager concluded in 1967:

> The procurement officer for the DX Project expressed astonishment to hear that Litton intended to bid on DX. He was frankly dubious of our abilities, especially our "systems" and "electronics" capabilities. As to Ingalls' shipyard management and production capabilities, we have had to make strenuous efforts in recent months to restore confidence of the Navy in us.[6]

Convinced by the memorandum, the Litton management began construction of a new shipyard on the west bank that would employ a "modular" method of shipbuilding, said to have been originated by the Swedes and the Japanese. Rather than building a vessel all in one place, from the bottom up, the envisioned assembly operation would construct parts that would be fitted together later.

"By using modular techniques the new shipyard incorporates the world's most advanced marine production technology," Litton said.[7] In March 1973, however, the U. S. comptroller general reported that the Ingalls subsidiary had largely abandoned its ambition for modular construction and had reverted to "nearly conventional" methods.[8]

Daring Venture

Litton announced to its shareholders in 1970 that in "building modern ship production facilities capable of achieving the highest possible efficiencies," the acquisitor had "invested more than $132 million."[9] That financing proves Litton's reputation as a master of money management. It had incurred less than

$3 million—spent entirely on design—of the cost of the shipyard. The state of Mississippi provided the funds, $130 million, by issuing bonds. According to the agreement, Litton leased the yard from the state but paid no rent for the first five years. In effect, Mississippi bore the rent obligation for Litton through 1972 by paying the interest on the bonds.

Litton had indeed chosen the more daring course (over the plan of mere facility repair)—daring for the state of Mississippi.

By mid-1970 Litton's investment was succeeding stupendously on paper. The navy decided on a large-scale fleet replacement and anticipated annual shipbuilding budgets of $3 billion for the decade. The most coveted prizes for shipbuilders were contracts for nine amphibious helicopter-carrying Landing Helicopter Assault ships (LHAs). Equally coveted were contracts for constructing thirty DD-963 Spruance-class destroyers. Litton persuaded the navy that its not-yet-operating yard was capable of the production and won both sets of contracts. The vessels were to be built over a period of years. Even under the original contract, the cost would exceed the navy's entire current annual budget. The destroyer assignment alone, Litton told its stockholders in 1970, "is the largest single contract in the annals of American Shipbuilding."

An irate senator from Maine claimed that the bidding and qualifications of a major competitor, Bath Iron Works, had been ignored. The award amounted to over-concentration in one shipyard, Margaret Chase Smith said. After the experience with Lockheed Aircraft Corporation and the C-5A, she wrote Secretary of the Navy John Chafee on April 24, 1970, "it is inconceivable to me that we would create a backlog larger than our annual ship-building budget in a single facility."[10]

The award amounted to favoring acquisitors and to a threat to competition, she said:

> As you know, throughout the McNamara era, large seg-
> ments of industry believed that the Government was really
> interested in awarding its major defense contracts to the
> huge conglomerate enterprises. Award of all of the DD-
> 963 ships to Litton Industries would, in my view, serve
> to confirm that this policy is being carried over into this
> Administration.
>
> The award would undoubtedly discourage medium-
> size companies from attempting to compete with these
> industrial giants for other major defense contracts, thus
> restricting competition by eliminating some of our most
> highly qualified contractors...

The attempt to create production capability in a new, untried shipyard by shifting to it more production than it could handle would cause atrophy in tried and proven yards, Mrs. Smith concluded:

> If the Navy fails to award at least some of these ships to
> Bath...in all probability the shipyard's destroyer capability
> will be seriously diminished. The loss of this valuable na-
> tional asset would significantly reduce the Navy's ability to
> obtain first-line competition for destroyer construction.

But Litton Executive Committee Chairman Glenn McDaniel had already described Ingalls' new yard to the navy. He described it the same way to the House Judiciary Committee. Again empha-sizing Litton's founding purpose of strengthening competition in industries it enters, he left no doubt that the unfinished yard would revolutionize world shipbuilding:

> Another distinguishing feature of Litton's history has been its ability to add new competitive strengths to companies it acquires. This is another implementation of the concept on which Litton was founded...
>
> An industry as moribund as any is the shipbuilding industry.
>
> Even as we entered that industry by the acquisition of Ingalls in 1961 we stated that it would have to undergo substantial innovative change...
>
> The new facility, the most advanced in the world, is now nearing completion.
>
> In competition with other large companies in the industry, Litton won the first major contract for serial ship production for the Navy. This new industrial asset will... be accorded world recognition for leading an old industry into a new era.[11]

Stockholders read:

> Litton's decisive innovation in American shipbuilding has been to unify engineering and production to design a ship not only for high performance but for economical production.[12]

100 % Price Increase

But the Judiciary Committee and the stockholders did not hear that the "economical production" of the LHA assault ships was spiraling more than 100% above agreed cost. Litton had secured the contract by promising to build the ships more efficiently than its competitors, but with the increase Litton's price far exceeded the competing offers that Litton's promise had eliminated.

As the price doubled, the delay increased. As of April 1972, the assault ships were two years behind schedule. The delays got worse.

Commercial shipbuilding contracts entitle the civilian customer to claims against the contractor for delay. *The Wall Street Journal,* June 30, 1972, reported:

> BEVERLY HILLS, Calif. Litton Industries Inc., beset by problems at its Pascagoula, Miss., shipyard, said it will pay $5.5 million to two ship lines, thus settling claims against it for construction charges and delays in the building of eight container ships.
>
> The conglomerate…said it reached agreement to pay $3.5 million to Farrell Lines Inc., New York, and $2 million to American President Lines Ltd., San Francisco, for "construction changes," excusable delays and "liquidated damage" related to the construction of the ships by Litton for the two lines.

But defense contracts—at least Litton's assault ship contracts—do not likewise entitle the government to claims for delay. In spite of the two year, ever-increasing delays, Litton claimed $270 million under those contracts alone *from* the navy, *in addition* to the more than doubled price. Over and above the price increases, Litton's *claims* against the navy amounted to $450 million. One half of the LHA claims alone far exceeded the cost of the new facility on which the vessels were to be built.

The status of the LHA production thus raised the prescient question the House Judiciary Committee had asked Litton's chief executive: whether an acquisitor might learn and develop its ship-building trade and profit in spite of malperformance not tolerated by commercial customers through securing U. S. government contracts. The ambiguity of Litton's naval contracts, the laxity of the navy's supervision of the construction, the navy's system of divided authority which beclouds responsibility for unwarranted contract awards, and the navy's protectiveness of a failing contractor gave the question new depth.

The House Committee on Armed Services met on April 17, 1972, to question navy officials about Ingalls' performance. Litton representatives had received an invitation to the closed session but did not appear. Chairman F. Edward Hebert presided:

> As you are aware, at our last meeting there was considerable discussion concerning reports that at least two of the Navy ship construction programs are in serious trouble. Specifically, the LHA and the DD-963 programs…
>
> These five ships [LHAs] are presently more than two years behind their production schedule, with very positive evidence that their costs, when delivered, will be substantially above the cost estimates…
>
> Stated simply, we have a very serious problem. [13]

Admiral Isaac C. Kidd, chief of Naval Material Command, said that Ingalls was entrusted with replacing ships collapsing from over-wear. The need was "catastrophic."[14]

He expected the delay of the LHA amphibious assault ship production to "impact" or delay the DD-963 destroyer production. Both types of vessels were to be built in the new West Yard. Work on the destroyers could not begin until the LHA's were finished and moved off the facilities. Nevertheless, the admiral said, Litton was confident that construction of the destroyers would begin as scheduled in January 1973.

To show its confidence, Litton had begun cutting metal for the destroyers in advance of schedule, on June 6, 1972. The future secretary of defense, Les Aspin, a member of the House Armed Services Committee, said the metal cutting served as a

> public relations gimmick that will divert already scarce workers from the LHA project which is only 2% complete. By stretching out the LHA program the Navy and Litton are only increasing costs and delays while hoping that the metal-cutting motions will give the appearance of an early start... [15]

A major cause of the delay and the doubling costs (exclusive of the $450 million claims) of the LHA production was the inexperience of the Litton managers transferred to Pascagoula. Among the problems were "overoptimism" and "repeated changes in top management personnel, who came [to Ingalls] largely from the aerospace industry and knew little of shipbuilding."[16]

When inquiring about the production, the Federal Maritime Administration received from Ingalls "irrelevant dissertations on what is done in the rocket and aircraft industries around the country." The FMA's reported: "If the contractor would refrain from using superfluous slogans and get on with the job, it would be better off."[17]

Severe labor problems beset the yard. The turnover of workers was double the normal rate of 30%. The modular assembly system worked poorly. The sections did not fit together. The contractor was learning at government expense, Admiral Kidd said:

> The contractor is learning that the management and technology planned for that mechanized shipyard is not adaptable. This learning process has produced errors, has caused delays and cost increases—not yet determinable by the Navy.[18]

$10 Million for Every Ship It Can't Build

Under the original LHA contract, Litton was to secure a profit of $10 million for each vessel. After the delays and cost increases began, the navy reduced the order from nine to five vessels. The navy had no other choice. Of course, then, the reduction was for the navy's "convenience," Litton asserted. Thus, the navy had to pay Litton the envisioned profit for the four never-to-be-built ships.

The "profit" amount paid was probably something less than $10 million per phantom ship. The exact profit figure was removed from the hearing transcript:

> Admiral KIDD. The profit was provided for at $10 million per ship.
> Congressman CLANCY. And if we terminated for convenience, there would be an estimated payment of [deleted] million of profit?
> Admiral KIDD. Correct.[19]

Had the navy declared the contractor to be in default of contract for delay, Litton could not have maintained that the ship order had been "terminated for [the navy's] convenience." The U. S. government would not then have been liable for the guaranteed profit.

The navy drafted lenient contracts. They did not specify the length of the contractor's delay that would amount to default. The navy's attorney at the hearing confirmed that the contract contained no default definition that would render the contractor's reward proportionate to his performance:

> Mr. PHELAN. I can't answer your question. Some technical man has got to say he [the shipbuilder] is so far behind no matter what he does he can't meet that delivery schedule. That is not a contractual legal question, that is a question for some expert to determine.[20]

The LHAs were costing more than twice the contracted price, were two years behind schedule, and were thus "impacting" the destroyer production, yet the navy was helpless to declare that its obligation under the contract to pay profits for canceled orders was in any way altered.

The navy nevertheless prepares formidable contracts, Litton contended. They are too "complicated."

> Congressman MONTGOMERY. I had the opportunity to meet with the president of Litton, Mr. Ash, and I got the impression from him that the Litton Industries was completely behind the shipbuilding program down there…He did say this, though, Mr. Chairman: Mr. Ash said, "It has gotten so you have such complicated contracts you spend more time working on the contracts, dealing with the Navy," than they do with engineering and researching. This has been the delay.[21]

In spite of the ambiguity of the contract, a contractor can maintain that he does incur a penalty for not progressing according to schedule. For each day of delay past the guaranteed delivery date, he is liable for a certain amount of money. The delivery date, however, is not specified in the contract and is determinable only by extensive negotiation. Even then, if the delivery date is established, the absolute maximum penalty for delay is minimal. Under the "impacted" DD-963 destroyer contract, for instance, it was less than one percent of the original contract price.

One reason that the contractor's penalty for delay is "infinitesimal," according to Admiral Kidd, is that the government, curiously, is liable for his insurance expense:

> Admiral KIDD. I started digging into it as to why we have such an infinitesimal penalty, and I talked to contractors. I asked them—not just shipbuilders—why don't we make it tougher on you? And the answer I got back was simple. "You can, but you will pay for it by the insurance rates that the contractors would have to take out." And in digging back through the files, I find indications that the Government insures itself, so this is about all we see in contracts of this sort. It is just a drop in the bucket, a slap on the wrist.[22]

The navy has been writing contracts for building ships since it supplied them for John Paul Jones. That after almost two centuries it fails to protect the public from uncontrolled cost and uncontrolled malperformance was, the navy's own spokesmen wholeheartedly agreed, inexcusable:

> Congressman PIRNIE. I gained an unmistakable impression this morning that the definition of "default," or the

"time for performance" was so indefinite that you were dependent upon cumulative evidence in order to declare a default…If we haven't learned how to draw a contract now that would cover these elements of performance, then there is something wrong…

But when you double the costs or when we have to pay twice as much for half the number of ships, there is something wrong and it is wrong in the contract, or else the company would not be continuing in its relationship with this project. Does the Admiral see what I mean?

Admiral KIDD. Oh, yes, indeed. I certainly do, Mr. Pirnie.

Mr. PIRNIE. You can understand the feeling of this committee as we go into this perpetual story of contractual misperformance, or nonperformance, without any apparent rights on the part of the Government in order to protect its interest, and then they begin to talk to us about prolonged litigation.[23]

Fifteen-Volume Claim Summary

Not just litigation of the navy's alleged liability for Litton's claims would be prolonged. Merely to comprehend the claims requires endurance. Though the government writes the contract, it holds its own feet, not the contractor's, to the fire. Trying to speed up Ingalls' construction process, the navy itself agreed to furnish certain equipment for the assault ships. But Litton asserted that delivery of the equipment had been late and thus the cause of additional expense. Also on the basis of "Navy interference in Design Development and over Management," Litton submitted a 15-volume,

6,000 page "summation" of the claims against the government of $270 million for the LHA construction alone. That amount was not included in the price increase from $113 million to $288 million for each LHA assault ship.

The only valid claims against the navy, Admiral Kidd told the Armed Services Committee, were from losses and damages incurred at Pascagoula by strikes and hurricanes. "But in my humble judgment," he explained, "those two elements would justify but a minute fraction of the total size of his [Litton's] claim."[24]

The "summation" of claims took Litton a year to prepare. To substantiate it, the company explained, would require another year. The navy received its copy of the "summation" only ten days before the hearing. The task of comprehending it had hardly begun, the navy spokesman said:

> A substantial amount of the contractor's presentation are pure predictions, his own predictions. We will not be able to comment on them with any degree of accuracy, I would say, for some time.[25]

Thus, the shipbuilder troubled by the purported complexity of his government contracts wrote a voluminous document himself. Earlier, in response to the question of whether Litton used a different procedure "in doing business with the Government than doing business with the private sector," Tex Thornton had complained: "The biggest difficulty of course is the paper work. It is tremendous doing business with the government."[26]

The selection of the Ingalls shipyard for construction of the assault ships involved a miscalculation of the required man hours by 200% to 300%. The estimate had risen $10 million per vessel before construction had even begun. Ingalls' capability to perform the contracts was doubtful. Yet the navy selected Ingalls for the largest shipbuilding award in history after a year-long "most comprehensive evaluation ever made by the Navy of a shipbuilding program."[27].

"How could you be so wrong?" the committee asked the admirals. The chief of Naval Material Command replied:

> We've got to remember that we made a decision on the basis of a shipyard that was nonexistent at the time, and a technique for production that was still sort of a gleam in somebody's eye.[28]

Explanations for selecting a nonexistent shipyard for the nation's most ambitious shipbuilding program over yards of proven ability range from patriotism to curious optimism over Litton's promises. Might the navy again award defense contracts, maybe in time of greater peril, on the basis of promised, not actual, capability? the congressmen asked.

> Congressman SPENCE. How do you think we got into this mess in the first place?

> Admiral KIDD. Overoptimism on the part of both the Government and the contractor in the beginning...

> Congressman STRATTON. Admiral, I am aware that you... were not here when the decision was made but the statement was an incredible one... when you said this contract was to a nonexistent shipyard to build a ship according to a technique that was also nonexistent. It is hard for me to see how anybody in his right mind could have thought this was a desirable thing to do.

Admiral KIDD. Well, if I may, I would take issue with that, Mr. Stratton, because others abroad and in the Soviet Union were having success, and are making it work, and making it work well. So I think the risk was well founded. As I said earlier, I have got that much pride in Americans' ability to build things, and I don't see any reason why—

Congressman STRATTON. There is no point arguing it now, Admiral. We don't have that ability, whether the Soviets do or don't.[29]

The failure of technology, facilities, management, and manpower brought on a failure of capital. Although Litton is to inject financial strength (besides managerial strength) into companies it acquires, Ingalls lacked sufficient funds to perform the contracts. Consequently, the navy made payments ahead of schedule to Litton, thus paying for work that was not performed. The amount advanced ranged from $100 million to $150 million by October 1, 1972. *The New York Times,* April 19, 1972, had reported:

> The House [Armed Services] Committee was told by Navy witnesses that Litton had already been paid 50 per cent of the contract price, whereas, even with the most generous calculation of engineering completed, only 25 per cent of the work had been done.
> "Litton is confronted with a critical cash flow problem," a Congressional source said. The company would not comment on these matters.

Under the terms of the contract, if the 24-month-delayed performance of the assault ships was not on schedule by September

1, 1972, Litton was to have repaid the advanced millions. When that time expired, the navy granted a six-month reprieve. When that extension expired on March 1, 1973, the navy granted another reprieve, extending the date to June 1, 1973. On announcement of the navy's second extension date, the acqisistor replied: "Litton believes such a repayment isn't due [even in June] and will oppose the Navy's claim."[30]

With that sword of Damocles hanging over the defense production, the navy's "on-site" inspector at Pascagoula, Admiral Charles N. Payne, turned to Ingalls' construction for *private customers* to find justification for the navy's "enthusiastic" acceptance of Litton's professed capability. He told the Armed Services Committee:

> I am Admiral Payne. I am supervisor of shipbuilding. I have been keeping up with the Farrell [Lines, Inc.] ships [under construction by Ingalls] on a regular basis with the Maritime Administration local representative down there. We have been very interested in the progress and the quality that the company [Ingalls] has been maintaining.
>
> My last reports indicate that although the progress is behind the original schedule and continues to be behind, and they are going to be delivering late, that the quality has improved tremendously on the last three ships. They did have quality problems on the first ship.
>
> The outfitting which is proceeding on the first ship, they are still learning on it, and it is somewhat below the standards in the industry, but they expect improvements in the second, third, and fourth ships as they are showing improvements in their construction.[31]

Whether Farrell Lines shared the inspecting admiral's good feelings for Litton's "learning" efforts was not certain. The delays and cost overruns did not abate. *Time,* July 3, 1972 (p. 32), reported:

> The Pascagoula plant is also far behind on construction of eight container ships for the Farrell and American President lines. Now scheduled for completion next fall, the first such vessel will be twenty-one months behind schedule and will cost about double its contract price of $21 million, making it the most expensive general cargo ship ever built. Litton will doubtless pay heavily for the overrun.

Litton might pay heavily, the article means, to its commercial customers. Not to its government customer. On June 30 Litton had agreed to pay $5.5 million in construction damages to the two maritime lines.

The cost overrun alone on the first ship of the Farrell contract would equal considerably more than one quarter of the entire construction cost of the liner Queen Elizabeth II.[32] Six months before the damages settlement, *Forbes,* December 15, 1971 (p. 18), reported:

> The Austral Envoy's deckhouse was placed on the hull structure, but the main deck sagged because beams underneath had been left out. Litton jacked up the deckhouse, put in the beams, and set down the deckhouse again—still off kilter. "It's only nine-sixteenths of an inch off," says O'Green [Litton's new manager of Ingalls]. "It doesn't affect the ship's performance."

Litton had transferred manager O'Green to Pascagoula from its nonshipbuilding Defense and Space Systems Group.

The Little Man Who Isn't There

A "lot of speculation, a lot of hypotheses that have been made of sand," Admiral Kidd stated, led authorities to believe Litton could build the ships. Who those authorities were and where they are now, was anyone's guess:

> Congressman RANDALL. Who were the people that made this decision? Who were the people that went so wrong? One of the problems, it seems to me, is that when we try to get at any of these things we are dealing with mush. The flag officers are transferred every two years. The Assistant Secretaries rotate every couple of years. We have a new Assistant Secretary who comes up with a new plan for controlling cost growth, and he is replaced, and somebody else comes in with a plan.[33]

In reply to the same prescient question asked two years earlier at House Judiciary Committee hearings—whether the frequent replacement of defense-contract administrators works to Litton's advantage—Tex Thornton said: "I don't think so. I never heard of that. No."[34]

The persons to whom Litton sold its undeveloped and failing shipbuilding systems were nowhere to be seen. None of the admirals in the Armed Services Committee hearing room could answer the question for which the hearing was called, "Who made the decision to retain Litton?" The agency responsible is the Office of Naval Materiel Command. Its chief, not surprisingly, assumed command *after* the decision:

Congressman RANDALL. There is no one in the room that really made the decision [to award the contracts to Litton] is that right?

Admiral KIDD. Made what?

Congressman RANDALL. The original decision, the important decision here. I am talking about back when the contract was let. Who was that?

Admiral KIDD. I don't know, Mr. Randall.

Congressman RANDALL. You don't know? Are you standing on that answer that you don't know who made the decisions?

Admiral KIDD. That is right. I wasn't here. I have no idea who made the decisions.[35]

But Chairman Hebert knew who made the decision:

After you have been around Washington as long as I have been, you will find out it [the official responsible for mistaken contract awards] is the little man who isn't there, he went away the day before.[36]

The navy admitted that the decision to award Litton the contracts was a disaster, being based on "a lot of hypotheses that have been made of sand." But at the start of the hearings the navy had attempted to defend the decision.

Admiral KIDD. Litton's Ingalls yard has not yet measured up even closely to some of the expectations predicted of it. However, I have enough confidence in the engineering capabilities of these United States of ours to be comfortable in the prospect that American ingenuity and production potential can be able to do as well or exceed performance abroad, as has been done in the assembly line of high volume production ships in the Soviet Union...

This contractor of ours is still learning to use his shipyard. He has a long way to go. This contractor is pioneering an exciting new approach in ship construction...We expected far too much of him at the outset, and the contractor was far and away overoptimistic in predicting his abilities to produce.[37]

Worst Indictment of a Shipyard

The admiral urged indulgence. He expressed hope for Litton's billion-dollar "learning" program and pleaded that Congress not fire Litton. Worse than Litton's failing performance, the navy counseled, would be to remove the destroyer production to a competitor's shipyard. The chief of Navy Material Command said:

There would, of course, be a possibility of having those ships built someplace else, if we find that the LHA impact on the 963 is too great. Moreover, if the Congress were to withhold funds for the...destroyers, this would remove a very important negotiating option from the hands of the Government as well as giving to Litton—mind you, giving to Litton cancellation charges which would be due Litton under the terms of the contract.[38]

Although Litton would be fired only for its inability to perform the destroyer contract, the taxpayers' liability for such a measure could approach a billion dollars. Litton could collect from $400 million to $950 million from the government if the navy canceled the contracts.[39]

Roy Ash resigned as president of Litton Industries in December 1972 to become director of the U. S. Office of Management and Budget. On his first day in that office, which determines federal expenditure, Mr. Ash assured reporters that by no means would he divorce himself from decisions affecting the navy:

> He plans to stress...sharper analysis of results in comparison with [government] costs. "We had better be sure we are getting our money's worth," Mr. Ash commented, because the federal government "is costing so much. We want to keep our eye on the expected results rather than just the best efforts to reach them," *The Wall Street Journal,* November 29, 1972, reported.

The budget office under Mr. Ash, the *Journal* continued, would "undertake a comprehensive examination of all government programs now in existence to determine whether they are actually meeting the purpose for which they were designated."

Navy and Maritime Administration auditors three weeks later found "poor workmanship and repetitive defects" in Litton's shipbuilding program. *The Washington Post,* December 19, 1972, reported the audit findings:

> Senior managers of the Litton Ship Systems Division "diluted" their effectiveness "because of the large amount of commuting" they had to do between Pascagoula and [other Litton operations in] California...

The planned allocation of manpower was inadequate for all ships.

The training program was "primarily subsidized by Government."

The company's own training program was "inadequate."

On December 20, 1972, the navy's director of Procurement Control, Gordon W. Rule, described the audit findings as "the worst indictment I ever heard of a shipyard"[40] (see p. 260). The Joint Economic Committee of Congress to which he spoke had invited Litton representatives to those same hearings at which the report was made public. Litton's Chief Executive Tex Thornton declined, however, replying: "a public discussion of the subject matter of these negotiations [of the acquisitor's claims against the navy] would be contrary to the best interest of both the government and Litton shareholders."[41]

Fixed Price Fiction

The navy's insistence that Congress allow Litton to remain as contractor at all costs may seem incongruous with Tex Thornton's assertion to the House Judiciary Committee in 1970 that a private contractor's performance of government contracts is a competitive process. He stated then:

> Mr. Chairman, in the Government business that we compete on, most of our competitors, 90 per cent or more of them are the giants, the big companies. It is though competition. More often than not, it is your exposure, your investment, your profits, and your return is less than it is on the commercial business.[42]

Bidding on government contracts may be a competitive process. Once a contractor wins the contract, however, as Litton has secured the assault ship and destroyer production, apparently it is assured of keeping the assignment regardless of how disappointing its performance and regardless of qualifications of competitors capable of finishing the project.

Mr. Thornton also averred that Litton's defense contracts were fixed in price, implying that Litton could not charge the government for price overruns:

> In the first place, most of our contracts for government work have been fixed priced. We have lost money on government contracts the same way that we have lost money on commercial projects. We operate under tight specifications at a fixed price.[43]

The LHA contracts were "fixed price." As seen, Litton's charges against the government for an overrun in excess of 100% of the contract price and its claims of over a quarter-billion dollars are only part of its 15-volume claim "summary."

A minute portion of the overrun consists of $7 million of navy funds that Litton misappropriated. That money, designated for the assault ship and destroyer production, Litton spent instead on its commercial projects.

$7 Million Misappropriation

In the *Report to the Joint Economic Committee of Congress on Controls Over Shipyard Costs and Procurement Practices of Litton Industries, Inc.*, March 23, 1972 (p. 16), the U.S. Comptroller General stated in blandest language:

The Defense Contract Audit Agency found that, during the period 1969 through 1971, Navy contracts for the LHA's and DD-963's were charged about $7 million for overhead expenses applicable to Litton's commercial work...

Although we did not review Litton's overhead-charging practices in detail, our selective examination indicated that they were resulting in the Navy contracts' bearing some of the overhead expenses applicable to the West Yard's *commercial* work [emphasis added].

The wrongfully allocated funds reduced the contractor's expenses on its commercial, nongovernmental shipbuilding. Thus, the misallocation amounted to the contractor's placing the navy's $7 million in his own pocket.

Litton is not inclined to return the millions, but the navy, the report happily assured, "has the matter under consideration":

The contractor believes that an adjustment should not be made for prior years' costs...Our limited review confirmed the DCAA finding that the contractor's method of charging costs incurred by Marine Technology [a Litton division] had resulted in Navy contracts' bearing certain overhead costs applicable to commercial work.[44]

Admiral Charles N. Payne, who spoke encouragingly of Ingalls' commercial production, supervised navy shipbuilding at Pascagoula. His office, the comptroller general's report states:

...is responsible for administering the contracts at the East and West Yards...It exercises surveillance over the contractor's operations to ensure conformance with contractual

requirements. To carry out this surveillance, [the office]...
had a staff of 275 civilians and nineteen military personnel.
This staff was involved in surveillance of such contractor
operations as quality assurance, planning, control of material procurement, and cost control.[45]

Cost control surveillance may well be a major obligation of that
staff. Congress, however, learned of the $7 million misappropriation
only because the Joint Economic Committee happened to request
that the comptroller general investigate the LHA and DD-963 construction. The full scope of misappropriations is anyone's guess.

Safeguards Flouted

A builder of ships who knows it can pass avoidable costs on to
an unquestioning customer has no incentive to avoid those costs.
Depending on the degree of the customer's laxity, the shipbuilder
might ignore federal laws enacted specifically to avoid needless expenses. He might even ignore contract requirements for proof of
claims against the complacent customer who orders the ships. That
a customer could be so obliging as to waive those requirements set
forth for his own protection was inconceivable.

A government contractor is unlikely to spurn safeguards against
unfair subcontract charges if he is held responsible for the cost overruns. If he does reject those safeguards he is apparently confident
that the government, unlike commercial customers, will bear the
unnecessary cost. According to the comptroller general's report,
Litton "had foregone safeguards for determining whether it was
paying fair prices":

> The purchase order files [of the East Yard] we examined,
> which covered larger buys, generally did not contain...

[data] for determining the reasonableness of subcontract prices.[46]

The Truth-in-Negotiations Act (76 Stat. 528) requires that contractors performing for the government negotiate with suppliers or subcontractors to secure reasonable prices. The report continues:

> The shipyard [East Yard] apparently did not hold negotiation discussions for most of the subcontract awards we reviewed. As a result, the lowest available subcontract prices may not have been obtained. The Truth-in-Negotiations Act provides that Government procurement officers hold such discussions.[47]

Examination of two hundred of the 224 total contracts revealed that the acquisitor had not complied with the act. Negotiations held in compliance with the act over the remaining twenty-four contracts resulted in substantial price reductions.

Litton's $270 million claim against the navy involving the assault ship production was based, as seen, on alleged costs incurred from the navy's alteration of plans, or "change orders," and from late deliveries of equipment. The comptroller general reported, however, that Litton's "budgeting and cost control systems" were inadequate to determine the cost of the change orders. The navy acquiesced to Litton, agreeing that "to segregate the costs of changes" so that claims may be verified would be impracticable.[48]

Litton's control system, the comptroller general found, was useless for verifying claims:

> The segregation of change-order costs, where feasible, is needed to provide a sound basis for negotiating change-order prices.[49]

Thus, by poor bookkeeping a contractor may thwart challenges to his claims and actually profit from his contract violations. A customer who accepts the violations should be ready also to accept a fifteen-volume claims "summation."

Enduring Mistakes

The West Yard's insufficient record keeping and cost analysis should have come as no surprise to the navy. Before the LHA assault ship and destroyer contracts were let, the navy itself had given the system in the East Yard a failing grade. The comptroller general reported:

> Approval of the contractor's purchasing system was withdrawn following a procurement review when the Navy determined that the purchasing manual did not fully implement the requirements of the Truth-in-Negotiations Act:
>
> -the bidders' lists were incomplete,
>
> -criteria for conducting negotiation discussions were needed,
>
> -procedures and capability for making cost analyses did not exist,
>
> -adequate documentation to enable reconstruction of purchase transactions was not present.[50]

More than a year later, after the award of the LHA ships and destroyer contracts, "the Navy found that most of the deficiencies previously disclosed had not been corrected."[51]

That the Litton headquarters contributes to the acquired subsidiary universal management ability adaptable to untried fields of endeavor may be, therefore, too sweeping a generalization. Safer to say, Litton bestows on the subsidiary a particular technique for securing government awards. That technique won for the acquired company the most colossal production contract the navy had ever granted—production to be performed by a "non-existent shipyard" with a "non-existent technology."

ITT secured anticompetitive advantage for its acquired subsidiaries through the mass of its acquired operations. Litton won that advantage for Ingalls through its technique for negotiating with government officials. Of these three acquisitors examined, the one that did not secure anticompetitive advantage has fared least well. LTV has undergone dissolution.

In April 2001, Northrop Grumman Corporation took over Ingalls Shipbuilding Corp., as well as Litton's remaining assets.

PART IV

✧ ✧ ✧

THE COUNTERATTACK IN DISARRAY

I've heard enough. Assistant Attorney General McLaren must have wiped his brow as he listened to the great acquisitors, Gulf & Western, LTV, Litton, and ITT, boast of those schemes for freedom from market rivalry. He resolved to continue the mission the Sherman Act began. To ensure that the market place is competitive, he would, quite logically, strive to ensure that competitors exist (see p. 18).

But government shelves already overflowed with regulations that purported to protect market rivalry. Statisticians, accountants, lawyers, economists, and staffs labored over them constantly. Mr. McLaren was convinced, however, that the very vastness of the regulations defeated their purpose—to ensure that the full public enjoys the complete force of America's economic capacity.

The collapse of America's fourth largest corporation at the time he prepared the ITT-Hartford suit more than convinced him of that defeat. The Penn Central bankruptcy showed either that regulations for protecting competition did not work, were not enforced if they did work, were evaded, or actually created danger.

If a physician were to minister to his patients only when the mood strikes him, he would fear for his practice. No such fear hovers over agencies created for ensuring square deals in the marketplace.

CHAPTER 11

The Interstate Commerce Commission—Surface Transportation Board

[A] most prolific source of financial disaster and complication to railroads in the past has been the desire and ability of railroad managers to engage in enterprises outside the legitimate operation of their railroads... The evil that results, first to the investing public, and finally, to the general public, cannot be corrected after the transaction has taken place; it can be easily and effectively prohibited.

> Report of the Interstate Commerce Commission's
> *Investigation of the New Haven Company; The New
> England Investigation,* 27 ICC 560 (1913).

The Penn Central Railroad succumbed from injury inflicted by its own hand. The injury was the drain of hundreds of millions of dollars incurred by its efforts to become an acquisitor. Had the Interstate Commerce Commission performed its most fundamental functions—and not thwarted basic public protection against financial manipulation—the largest bankruptcy then in history could not have occurred.

By 1970, the Penn Central controlled over 190 corporations. More than eighty of them operated outside the field of transportation. Its real estate holdings ranged from the Six Flags Over Texas amusement park to the Waldorf Astoria Hotel. Consolidated assets surpassed $6.85 billion.

By the early 1960s, plotting diverse takeovers had become more thrilling to the Pennsylvania Railroad's management and board of directors than running a railroad. Stuart T. Saunders, chairman of the board and chief executive, had been in the railroad business for almost his entire working career. He was getting tired of it, he said. His specialty, moreover, was not railroad operations but finance and public relations—especially stockholder relations.

He took control of the Pennsylvania Railroad in 1963 and by 1966 had increased its dividends from less than $7 million a year to more than $30 million.[1] Revenue increased also, but not that much. Part of Saunders' management technique was to issue stern orders to cut costs in any way. His subordinates obeyed. The railroad's steady decline in service proved their obedience.

By 1965, the Pennsylvania Railroad management and board had abdicated control by giving Finance Director David Bevan and Treasurer William Gerstnecker complete freedom to dispose of assets necessary for railroad operations in order to acquire companies far afield from railroading. The two officers gladly spent millions not only without the board's approval but without its knowledge.[2]

Funds for roadbed maintenance grew scarcer. Trains had to creep over long stretches at speeds that would have been slow in the 1880s just to stay on the wobbling rails. Still, derailments were pandemic.

Not surprisingly, in 1968 the Pennsylvania Railroad was enthusiastic over an exception to its diversification program—merger with another railroad, the efficient and well-managed New York Central. The courts and the regulatory agencies, swayed by the old argument that combining of operations and management would somehow enhance efficiency, permitted the merger—then the largest ever—as an exception to the Celler-Kefauver Amendment which, as seen, prohibits combination of enterprises engaged in the same industry.

A prime consideration that induced the capable New York Central management to agree to merger was the idea that trains of one company could run over the rails of the other.[3]

The argument for the combination, facile as always, supposed that certain duplicate stretches of tracks of both companies could be eliminated. However, the Pennsylvania Railroad tracks, which hardly supported trains of one railroad and in some places not at all, held up no better under the weight of two railroads.

Not on Same Track

Nor did the combination of managements produce economy of scale. The New York Central officers, more dedicated to the operation of railroads, didn't like losing control of revenue their carefully maintained assets generated. They didn't like to see it spent on activities impertinent to them:

> As the railroad operations deteriorated, the philosophical split already in being between the heads of the Penn Central's component railroads became virulent. Alfred E. Perlman, who had been president of the New York Central, did not believe in diversification [by purchasing other corporations]. Even before the merger, Perlman loved to show visitors technological improvements he was making on the railroad. He would guide them through every part of a new electronic switching yard, for example, explaining how it would improve service for shippers while reducing costs so much that the Central would get its money back quickly.

Then, in a slap at the Pennsylvania, he would say, "I'm putting every cent I can find into making this a better railroad. I'm not putting it into some silly amusement park" (a dig at the Pennsylvania's investment in the Great Southwest Corp. and the Six Flags Over Texas park).[4]

The computer systems of the two companies could not be synchronized. Entire trains were lost for weeks and longer. The cost of replacing the two computer systems—both entirely adequate prior to the attempt to economize through combining—surpassed $20 million.

Particularly galling to the former New York Central managers was the $21 million spent to buy Executive Jet Aviation. The Pennsylvania Railroad's board had approved Finance Director Bevan's and Treasurer Gerstnecker's purchase of the company even though the railroad's control of an air carrier was no doubt illegal. The purchase violated the Federal Aviation Act. The Civil Aeronautics Board was certain to disallow it.[5]

Department Store of Transportation

The decision to enter the air carrier industry unlawfully was only the beginning of the railroad's adventure into diversification for diversification's sake. Whether the companies acquired (for diversification) would return any income to the railroad appeared to be no concern of the Pennsyvania board. The stockholders would bear the loss. And because of the failure of the regulatory agencies to administer legislation designed to protect investors, the stockholders would never know of their jeopardy until too late.

The First National City Bank of New York—eager for the millions the railroad would pay it as interest—led the way in providing easy and ample credit. Like Chase Manhattan, it was not averse to helping acquisitors grow.[6]

The bankers loaned their depositors' money as quickly as the railroad chose what companies to buy. Rather than investigating the investments, the bankers dwelled on the railroad's reputation of invincibility. The loans themselves struck down the venerable reputation that gave them life. Unable to meet its financial obligations in mid-1970, the Penn Central could not secure loans to avoid bankruptcy because its credit was already exhausted by the diversification loans.

Bevan and Gerstnecker enjoyed complete power to obligate the railroad for Executive Jet Aviation's purchases of aircraft—power as illegal as it was unbridled. Together with the president of Executive Jet Aviation, Brig. Gen. Olbert F. Lassiter (U. S. Air Force, ret.), they purchased two 707s and two 727s from Boeing at a cost of $26.2 million.[7] The First National City Bank of New York made plain that it considered the borrowing to be the obligation of Penn Central.

General Lassiter, even more confident than Bevan and Gerstnecker of the infinite depth of the railroad's coffers, signed a letter of intent with Lockheed for delivery of its L-5000, the civilian equivalent of the C-5A military transport, the most gigantic airplane ever built. He wanted six of them, he told Lockheed, for a total cost of $136.5 million. Boeing delivered. Fortunately, Lockheed did not.

To make use of its new jet fleet, Executive Jet Aviation needed to acquire customers. It decided to do just that—acquire by purchase a worldwide network of airlines, just as it had purchased the Boeing jets. Executive Jet Aviation would then order the airlines it owned to lease its aircraft. Negotiations for the purchase of air transport companies were conducted in France, Germany, Indonesia, the

Netherlands, Panama, Saudi Arabia, Spain, and Switzerland. Two purchases were Transavia Holland and International Air Bahamas, which flew between Nassau and Luxembourg. The transactions were so obscure, and Bevan's use of Penn Central funds for purchase so free-wheeling, that $4 million transferred to a Liechtenstein shell corporation was never seen again. Where those millions went no one knows.

President Stuart Saunders explained in a public address early in 1967 that the railroad, by branching into the air transport business, intended to become "a department store of transportation."[8] He did not elaborate that for a railroad to control an air carrier is illegal, nor that the millions of dollars of stockholders' money drained into it would not be seen again. After the address, the railroad's public relations director explained that Executive Jet Aviation, with its acquired airlines, was the instrument for the diversification Saunders had described. Immediately Treasurer Gerstnecker reprimanded the director. The railroad's control of the air carrier was not to be mentioned in public, the treasurer insisted, because the Civil Aeronautics Board might be listening.

The board did find out. Penn Central tried to hide its ownership of Executive Jet Aviation through sham sales agreements (in which Penn Central tried to retain the right to repurchase the air carrier) with United States Steel and Burlington Industries. But the Civil Aeronautics Board ordered Penn Central to liquidate its holdings in Executive Jet Aviation. The order found thirteen separate violations of the Federal Aviation Act and levied fines on the air carrier and on Penn Central—slightly smaller than the largest penalty ever imposed by the board.[9]

The management that incurred those fines never found a purchaser for Executive Jet Aviation. The company never recovered $1 million of its $21 million investment.

Too Many Passengers

That amount, of course, was only part of the diversification-by-acquisition cost. Two hundred nine million dollars was spent directly on acquisitions. Millions more were distributed by Saunders through excessive dividend payments in efforts to boost the price of Penn Central stock so that it could continue to buy other companies.

Speaking as audaciously as he had spent, David Bevan told the Senate Commerce Committee on August 6, 1970, that income from the companies outside the rail industry that Penn Central had acquired enabled it "to keep the railroad running":

> In summary I believe our financial management over the years has been good. Even with all the adverse circumstances I have outlined, we were able to produce the money necessary for the operating people to keep the railroad running in the face of deficits and that was no small job.
>
> I might add that it would not have been possible without the income made available through our new diversification programs...Those dividends and income from other non-railroad properties have served to blunt the losses from passenger service and have provided the margin necessary for continued operation of the Railroad. In other words, our investment in non-railroad companies yielded a much better return than the Railroad itself, which would have been in much more serious trouble without the benefit of diversification.[10]

That the railroad profited at all from the companies acquired after the beginning of the diversification program in 1963 is false. The income those companies generated never equaled the interest the railroad had to pay on the money borrowed to acquire them.

Thus, the return from the $209 million of capital directly invested in diversification-by-acquisition between 1963 and the time of bankruptcy in 1970 is zero.

Finance Director Bevan's placing the blame for bankruptcy on the alleged doldrums of the railroad industry and his statement that the "non-railroad companies yielded a much better return than the Railroad" are insupportable.

One Penn Central official remarked that if the $25 million used to purchase the Great Southwest Corporation, a real estate developer, had been used to refurbish railroad yards and tracks, the investment would have paid for itself annually three times over. Invested in Great Southwest, the $25 million yielded three percent, not three hundred percent.[11]

On January 11, 1967, EJA President Lassiter wrote to Finance Director Bevan in appreciation for the railroad's financing of the jet aircraft purchase. "Thank you," he said, "for laying your career on the line."[12] His words were more apt than he meant. Five years later, Bevan, Lassiter himself, and Charles F. Hodge (a senior member of the investment firm of F. I. Dupont-Glore Forgan, which plotted the railroad's diversification program), were indicted in Philadelphia for having conspired to drain "substantially the resources of the Penn Central, contributing to its bankruptcy in June 1970."[13]

Costly Appearance

Penn Central's rush to self-destruction should have been obvious to the public. Why its disastrous capital outflow in the late 1960s passed unnoticed is largely attributable to Saunders' use of misleading bookkeeping practices accepted by the Interstate Commerce Commission, the Securities and Exchange Commission, and the accounting profession. And his payment of exorbitant dividends did not inspire stockholders to inquire about operations.

The deception was for two purposes. First, stockholders would be lulled into a false sense that all was well. They would be less likely to question the claim that diversification was profitable and certainly less likely to depose the management (as the company did when it finally—bankruptcy pending—looked at the facts). Second, by reporting false profits and paying high dividends—even while more cash flowed out of the company than in—Saunders and Bevan could inflate the value of Penn Central stock. So inflated, the stock of course could buy more companies.

In 1968 Penn Central fortified public confidence in its operations by reporting a loss of more than $20 million as an $88 million profit. The next year the management told the shareholders it was sorry to report that the profit from operations had decreased to $4.4 million. The management did not say that it was sparing them some details of the sorrowful story. Rather than earning $4.4 million as claimed, the company had lost at least $90 million—possibly more than $110 million.

A two-edged sword for such reporting of inflated, or rather nonexistent, profits is the habit of adding to the income figure amounts that should not be added and of failing to subtract amounts that should be subtracted—a familiar sword. Gulf & Western used it partially by adding to operating earnings its nonrecurring stock market gains.

The method is especially valuable to a company that acquires other corporations because investors receive an exaggerated notion of the profitability of the acquirer's operations and mistakenly believe that the extraordinary gains will recur. They place a higher value on the acquirer's stock, thus providing the acquirer with more paper wealth with which to purchase other companies.

Penn Central failed to deduct from its income figure vast sums paid in wages and the cost of depreciation of machinery—amounts that are no part of net income—and thereby reported profits it had not earned. Charging equipment repair costs to capital rather than to operating expenses (that should be deducted from income) alone inflated reported income by more than $20 million.

> Such normal operating expenses as wages and phase-out costs were taken out of capital rather than income, thus inflating Penn Central's reported profits by millions of dollars. Sales of real estate, which should have been designated as extraordinary items, were credited to normal income, as were the profits from securities sales…In some transactions the railroad company sold assets to its financial subsidiary at many times their book value and credited the difference to its own income profit.[14]

The ambitious railroad accountants devised imaginative techniques. While Penn Central revenue declined, the salaries of its executives rose sharply. Charles Hill, manager, general accounting, received a salary increase of 100% between October 1967 and September 1969. The recommendation to Finance Director Bevan for Hill's salary increase of October 1967 states:

> He is extremely creative, is an excellent manager and is very cost conscious. *His imaginative accounting is adding millions of dollars annually to our reported net income* [emphasis added].[15]

No Income—Dividends Galore

The Penn Central income figure given the public was further inflated by supposed dividend payments to Penn Central from acquired companies it controlled. Dividends are to come from a company's earnings. The payments, however, far exceeded the subsidiaries' income and came instead from capital resources those payments depleted. Dispatch Shops, Inc., earned less than $3 million in 1969, but Penn Central took from it a dividend of $4.7 million. New York Central Transport Co. earned $4.2 million in 1969 but delivered to its parent company a dividend of $14.5 million and had hardly any assets left. Manor Real Estate suffered a $7 million loss in 1969, but Penn Central took from it a $2 million dividend.

Depletion of the acquired companies' resources of course impaired their earning capacities. With lower earnings or none at all, their alleged dividends to their parent's headquarters could come again only from their dwindling capital assets. The vicious cycle could not continue because soon the subsidiaries had no more assets to give Penn Central under the guise of dividends.

The parent company, in turn, continued to pay supposed dividends in 1968 and 1969 to its stockholders, who thought they were receiving a distribution of income from operations, even though in those two years the company lost over $110 million from its operations. Saunders staunchly insisted that the railroad's appearance of profitability be maintained at all costs. Even Bevan advised him that the railroad, desperately short of capital, had exhausted all sources of resupply.[16]

Saunders had the answer: If money is so scarce, borrow it. Bookkeeping legerdemain still prevented disclosure to investors of the company's deterioration, and Bevan managed to borrow $700 million in 1968 and 1969—some amounts at interest exceeding 10%. In the best of times the company's assets did not produce income at half that rate. Thus, Penn Central could not pay the interest, let alone the principal, without continued borrowing.

In April 1970 Bevan decided to sell bonds to the public (to be issued as a debt of the investment company subsidiary, the

Pennsylvania Company). But by then every asset Penn Central owned was mortgaged. Investors rejected the bond offering, thus pulling down the scaffolding of Saunder's costly accounting gimmickry. Now the public knew the company could not pay its debts. At last, the board of directors had to admit that the company was folding.

"Penn Central board chairman Stuart Saunders vigorously defended the solvency of his company today at the annual stockholders' meeting here," *TheWashington Post,* May 13, 1970, reported from Philadelphia. He admitted that the company could not continue to lose money at the rate it had been. But as usual, the scapegoat was the obligation to provide passenger service, "a dominant factor for keeping our railroad in the red." But Congress might relieve the company of that duty by creating a government corporation to run the passenger trains, Saunders told the stockholders. "To emphasize the urgency Saunders feels over rail passenger operations, he asked all stockholders today to contact immediately members of the House Interstate & Foreign Commerce Committee, headed by Representative Harley 0. Staggers and to urge passage of the rail transportation bill," the *Post* concluded.

Finally, the board of directors was worried. "Why can we no longer sell bonds even at interest above 10%?" the directors asked. Undaunted, that same month the board voted Saunders a salary increase—and to other officers as well.

The public stockholders bore the expense of the management's diversification and acquisition programs. Their per share value was to decrease from $86 in 1968 to $5 and less after bankruptcy. Now the management, unable to borrow from regular sources, conceived the idea of keeping the company afloat by spreading the loss to an even greater public: the entire United States citizenry.

Bankruptcy Victory

In May 1970 Saunders called on Secretary of the Treasury David Kennedy and Secretary of Transportation John Volpe to ask that the government come to the rescue by providing a loan guarantee under the Defense Production Act of 1950 (50 U.S.C. App. 2091). The purpose of the act is to prevent the collapse of corporations that are vital to the nation's defense. In such an emergency, the government is authorized to guarantee loans of private lenders to distressed corporations in defense industries. The guarantees are called Victory Loans.

Late that month the Department of Treasury outlined to an assemblage of 122 officials representing more than seventy banks its plan for granting Victory Loans to Penn Central. The government would guarantee the railroad's repayment of $225 million of loans from the banks. So as to fit the terms of the Defense Production Act, the Pentagon would act as the guarantor agency. More specifically, the Department of the Navy would shoulder the obligation.

Most any large corporation, and certainly the nation's sixth largest, is bound to have some ties to defense production. If Penn Central was eligible for rescue from bankruptcy by the Pentagon, then almost any defunct corporation that meets a test of size is entitled to Victory Loans. By that reasoning, any corporate management, no matter how derelict or reckless can be assured of government salvation so long as its operations are adequately gigantic.

The Department of Justice went a step beyond that reasoning by determining that Penn Central qualified for Defense Department assistance even though bankruptcy would not interrupt functions allegedly vital to national security—i.e., operation of the railroad. Assistant Attorney General William H. Rehnquist, soon to be appointed to the U. S. Supreme Court, informed Deputy Attorney General Richard G. Kleindienst that "insolvency is most unlikely to lead immediately to a cessation or to any serious curtailment of services." Nevertheless, he concluded, the United States "is authorized by the Defense Production Act to make the guarantee in question."[17] If the company then losing $700,000 a day could not pay the $225 million loan, as it admittedly could not unless another, much greater loan was forthcoming, U. S. taxpayers would pay.

The management congratulated itself for having ignored the old rules and for having found still another device for averting doom. Using railroad resources to purchase other companies, depleting those companies' assets, resorting to artful accounting techniques—all had only obscured the coming of Armageddon. Now, at last, a truly infallible device provided by the good will of all America would allow the Penn Central managers once and for all to avoid the consequences of their efforts to turn the railroad into a diversified acquisitor.

But this time the managers' sense of relief was short-lived. Chairman Wright Patman of the House Banking and Currency Committee, a framer of the Defense Production Act, disagreed with the Department of Justice's generous interpretation. By no stretch of the language was Penn Central eligible for Victory Loans, he contended: "The Act was never meant to prevent the insolvency of a large corporation only tangentially involved in defense contracts, but rather was intended as a way for small and medium-size contractors to expand their productive capacities so as to be able to meet critical defense production needs."[18]

The railroad managers flew in the company airplane from Philadelphia to Washington (their regular conveyance for the trip) to present their case face-to-face to Mr. Patman. He drew out from them that a $225 million Victory Loan would tide the company over only until October—i.e, for four months. Then more multimillion-dollar loans would become due. If the government did not guarantee loans of $500 million in October, the first $225 million guaranteed by the taxpayers would be lost and paid by them. Mr. Patman pledged his opposition.

But any firm as gigantic as Penn Central obviously had tentacles locked around the economy, Federal Reserve Board Chairman Arthur F. Burns said. Presaging the winning arguments of ITT that size exempts an acquirer from laws of the United States and of his successors that American International Group is too titanic to let sink (see chapter 14), Chairman Burns urged that the government guarantee the loans, that Penn Central be kept afloat. But the incoming director of the Office of Management and Budget, George P. Shultz, former secretary of labor, later secretary of treasury and secretary of state, agreed with Mr. Patman. A firm should not hoist itself by its tentacles above the market the "invisible hand" sets in order—let the chips fall where they may, he said.[19]

For a while the administration continued to proceed with the bailout scheme. But on Friday, June 19, 1970, opposition mounting, the Justice Department abandoned the attempt. The company could not pay its creditors. They would demand liquidation. To keep the railroad intact, the only track still open was to petition for bankruptcy. At 5:30 Sunday afternoon, hours before insurmountable loans would become due, Federal Judge William Kraft in Philadelphia signed the order for bankruptcy and rendered official the greatest corporate failure of all time.

Like the sinking of the *Titanic*, the collapse of Penn Central destroyed confidence in the judgment of experts. Reliance on enforcement of safeguards thought to be infallible, the best modern mind could devise, brought disaster.

The railroad industry has the longest history of government regulation. Except for gas and electric utilities, no industry reports as much data to the government. *I don't have to worry—I know I have the best management possible,* the average purchaser of Penn Central stock would have thought on inspecting the vast array of statutes imposing duties of the Interstate Commerce Commission and the Securities and Exchange Commission.

The Penn Central managers, however, were free to dissipate railroad resources on whimsical acquisition schemes—and to conceal from investors the calamity those schemes wrought—because regulatory agencies chose not to act in two areas of broadest responsibility.

First, the regulators not only shirked duty to *provide* protection to the public but *cut off* protection it already had. The Interstate Commerce Commission issued no regulations under its clear authority to require railroads to disclose their financial condition upon sale of stock, bonds, and other securities to the public. And merely by classifying railroad holding companies as "carriers," it *severed* shareholders' rights to protection against financial manipulation.

Second, the ICC did not bother to flash red lights at madcap merger crossings. Congress granted the ICC power to prohibit those mergers in 1920, after the commission itself urgently requested it in 1913.

I. Secrecy

The Penn Central managers escaped Securities and Exchange Commission supervision. The ICC's assertion of unexercised authority over the Penn Central's nonrailroad (as well as railroad) properties nullified the SEC's power to reveal operating conditions and uses made of company assets. Had either of the two agencies disclosed, as Congress intended, Penn Central's extravagance and deterioration, the managers could not have continued their disastrous acquisition course to the utter end. Because the public did not know the facts, those managers remained at the throttle even after the wheels of the great engine lay gaping at the sky.

On September 24, 1970, the House Commerce Committee asked why the ICC failed to signal the hazard:

> Congressman MACDONALD (presiding)...Despite the Congressional mandate 83 years ago that the [Interstate Commerce] Commission embark on a program that would establish an efficient and economically viable rail system in this country, the rail system in this Nation is neither efficient nor economically viable. There is strong evidence of misrepresentations to the public through incomplete information prospectuses filed with the ICC.[20]

At issue was the commission's duty to require railroads to disclose their true financial and operating condition so that the public will not be misled when investing in railroad securities.

Safety Never Provided

ICC representatives at the hearing first insisted that the commission requires railroads to specify in any printed offer (or prospectus) of public sale of securities the very same information that the Securities Act of 1933 (48 Stat. 881) requires of other corporations. Schedule A of that act sets forth thirty-two fundamental requirements—including a certified balance sheet, disclosure of management compensation, and management's interest in the transaction advertised—necessary for informed investment decisions. Congress considered those thirty-two requisites to be so minimal that any security issuance without them must be driven from commerce. The thirty-two requisites are merely "what competent bankers require from borrowers."[21]

But the Securities Act of 1933 does not apply to railroads. The ICC already had the authority to require the thirty-two requisites from carriers. No need to give the ICC power it already has, Congress thought. But after the bankruptcy, it found that the ICC did not require any of the thirty-two items.

The ICC explained to the investigating committee that it nevertheless enforces the disclosure standard demanded by the Securities Act even though the commissioners have not set forth that standard in regulations.[22] ICC Supervising Attorney John M. Mattras said at the hearing:

> We require prospectuses [describing securities that railroads offer for public sale] to furnish substantially the same information as that required by the Securities & Exchange Commission [pursuant to the Securities Act of 1933 which that agency administers].[23]

The investigators hastened to test that assertion by examining eight railroad prospectuses approved by the ICC.[24] None of those documents selected by the commission satisfied the disclosure standard the Securities Act required. The regulatory agency's claim that it enforces compliance with the act was disproved.[25]

With no uniform rule for disclosure, the ICC required of different carriers differing standards of disclosure. A carrier's lobbyist, therefore, could plead for leniency. He would not have been told that any uniform standard prevented special treatment.

Absolute Surprise

As with enactment of the Securities Act of 1933, Congress largely excluded railroads from the protection that later legislation affords because the ICC already had power to apply to interstate carriers the safeguards the statutes provide investors in other industries.[26] It assumed that once it established the minimum standards for investors' protection generally (by the passage of those acts) the ICC would rush to impose the same standards on railroad investments. But the ICC was asleep at the switch and stayed asleep.

The Securities Exchange Act requires disclosure of possible conflicts of interest between management and stockholders. Enforcement of the act to railroads, again, was left to the ICC.[27] But the Penn Central reports filed pursuant to the act contained no data sufficient even to alert ICC officials to the railroad's accelerating deterioration.[28] The ICC chairman himself was "taken by surprise" by the bankruptcy.[29] Chairman Dale W. Hardin's duty was to examine the railroad financial data his commission collected. Since it did not alert even him when rumors of collapse were rampant, the House Banking and Currency Committee asked what use the data served:

> On December 20, 1971, eighteen months after the failure of the railroad, the ICC issued its "definitive study" of the Penn Central collapse. On the basis of this 1,760 pages of rehash, one would think that the ICC was a branch of the National Archives rather than a regulatory agency charged with the responsibility for seeing that the railroad industry is operated in the public interest. While the ICC's study might serve a useful purpose to historians, it is of little significance to the thousands of investors who lost their holdings because the ICC failed to perform its primary function of regulating the railroad industry…Why didn't the ICC analyze these documents?[30]

The Trust Indenture Act of 1939 protects small investors who make loans to corporations. The loans are often too small to warrant the individual creditor's (or debenture holder's) expense of suing the corporation if it does not repay the loan. Thus, the practice developed for many small creditors who aggregately loaned a large sum to a corporation (by purchasing its debentures) to select a trustee, such as a bank, which would bring one suit against the borrower on behalf of them all.

The trustees, however, were not always vigorous in asserting the creditors' rights. Usually the creditors were dispersed over the country and had no way to know whether their trustee was conscientious, or whether—often the case—the trustee's interests conflicted with their own. Indenture trustees often connived with defaulting corporations to defraud the creditors they were dutybound to protect. Again prompted by the calamity of the 1930s, Congress in 1939 enacted the Trust Indenture Act, to prevent the trustees and large debtors from colluding against the small creditors. But again it excluded the railroad industry from the act because the ICC already had the authority to extend protection in that area. The act renders illegal those relationships between indenture trustees and corporate borrowers that might conflict with the creditors' interests.

But Penn Central, already lurching off the rails, sought ICC sanction of just such a relationship. In April 1970 it submitted for approval a trust indenture for borrowing $100 million from the public through its investment company subsidiary, the Pennsylvania Company. The trustee was to be Manufacturers Hanover Trust Company—itself a creditor of Penn Central and its subsidiaries in the amount of $33.2 million. Section 310 (b) of the Trust Indenture Act proscribes exactly that relationship, for the trustee and the debenture holder (whom the trustee owes the duty to protect) would, in the event of the borrower's bankruptcy, wrangle over the same assets. Manufacturers Hanover Trust did take for itself the bank accounts of the Penn Central subsidiary—assets that the debenture holders would have fought for.

The conflict never arose because the public wisely rejected the subsidiary's $100 million debt offering. Nevertheless, the proposed indenture is the exact peril to investors the Pennsylvania Company expected the ICC to approve. The ICC did not show that it would have rejected the borrowing.

Safety Severed

Impairment of the congressional design for the security of the investing public has resulted only, to this point, from the ICC's nonfeasance. The passage of the Investment Company Act of 1940, however, opens a new panorama of regulatory futility. Then the ICC began affirmative action to short-circuit lines of protection specifically designed for the public investor—protection he would have received but for the existence of the ICC.

The Investment Company Act protects investors who purchase the securities of companies that exist to invest in securities of other corporations. Congress passed the act in 1940 because manipulation of the public's $7 billion investment in investment companies had been rampant. The public lost over half that amount.[31] Transactions between an investment company and the corporation in which it invests—particularly transfers of assets, loans, and loan guarantees—are to be reported to the Securities and Exchange Commission, which enforces the act. The act seeks to prevent any investment company from acquiring excessive debt, resorting to unsound accounting practices, or allowing more than half its board to be bank officials.

150 Years to Pay

The Investment Company Act (sec. 1[b]) sets forth eight safeguards. The railroad's subsidiary investment company, the Pennsylvania Company, violated all eight.

The $100 million attempted debt offering betrayed the Pennsylvania Company's (operated by the Penn Central managers) threat to public safety. The investment company was to shift $84.1 million of the proceeds from the offering over to Penn Central in exchange for nonrailroad properties in which Penn Central had invested no more than $29.3 million. On the basis of their 1969 earnings, it would have taken those properties 150 years to have earned back their $84.1 million purchase cost.[32] Thus, any assertion that the transfer was in the interest of the investment company subsidiary—shares of which were traded to the public on the New York Stock Exchange—is incomprehensible.

The Investment Company Act, then, is another part of the congressional plan to provide investors with protection that would have revealed the railroad acquisitor's mismanagement when disaster course was still reversible.

A few days after the filing of the bankruptcy petition, and after 150,000 public stockholders had lost 95% of their Penn Central investment, Harley O. Staggers, chairman of the House Commerce Committee, inquired of the Securities and Exchange Commission why it had not required the Penn Central's investment company to adhere to any of those eight safeguards.

Chairman of the SEC, Hamer H. Budge, replied two weeks later that the question was interesting but not one he could answer right away.[33]

On September 16, 1970, six weeks after the House Commerce Committee had asked, it heard that the Investment Company Act "appears" to be inapplicable to the subsidiary investment company.

The ICC had ordered that the Pennsylvania Company be classified as a "carrier," the SEC explained, because the company controlled a railroad (the Wabash), although more than half the investment company's assets were invested in noncarriers.[34] That order, the ICC declared, rendered the company subject to ICC regulation. Because the Investment Company Act excludes from its provisions any company subject to the ICC, the Securities and Exchange Commission could not apply the act's protection to the Pennsylvania Company.

Hence, the management of any investment company, mutual fund, or other enterprise that holds securities for the purpose of investment can deprive its stockholders of protection that Congress declares to be essential and absolutely minimal, if the company or fund purchases 51% of the stock (controlling interest) of a railroad—however small that stock's proportion to total securities held.

When writing the Investment Company Act exclusions, Congress never supposed that such railroad investment would switch off investor protection. It excluded ICC-regulated companies from securities legislation because, as usual, it assumed that the ICC with its existing power would impose equivalent regulation. The ICC contended, however, that it has no such power, and that its affirmative orders can only take away and not replace provisions that guard the public from financial manipulation. That reasoning, disproved by the fact that the commission is empowered to authorize securities issuances (54 Stat. 907, Sec. 20(a)), defeats the major purpose of the commission's existence.

Other holding companies owned by acquisitors also have freed themselves of the act's minimal standards by mere purchase (even part purchase) of a carrier. International Utilities Corp. acquired Ryder Truck Lines in 1965. The SEC then promptly declared that the entire corporation (represented on the Penn Central board of directors) was exempt from the Investment Company Act's restraints against financial manipulation.[35]

The ICC allowed investment companies to sever shareholders' rights under the act by purchasing a carrier even though that purchase violates the agency's enabling statute, the Interstate Commerce Act. By virtue of such an illegal purchase, the Alleghany Corporation, acquirer of the nation's largest mutual fund, Investors Diversified Services (with assets of $6 billion), has removed itself from SEC supervision. In 1968, Alleghany purchased Jones Motor Co. The ICC itself determined that the purchase violated Sec. 5(4) of the Interstate Commerce Act. The commission nevertheless rewarded Alleghany with the prize it sought—freedom from the Investment Company Act's protection of widows and orphans. Jones Motor Co. constituted no more than 12.5% of the total assets of publicly held Alleghany.[36]

The more audacious the violation, the greater the chance for its success (see p. 3). After having approved Alleghany's purchase of Jones Motor Co., the ICC permitted Greyhound Lines' diversification transactions although the commission's staff found them to be "not consistent with the public interest":

> Greyhound Lines' working capital is depleted by advances to affiliates. Its bus equipment is encumbered to permit it to expand at a greater rate into non-carrier investments. We find this situation not to be in the interest of Greyhound Lines, since it could in a financial squeeze seriously impair the carrier's ability to perform its service.[37]

The transactions that caused the capital depletion and jeopardized Greyhound's public service violated ICC regulations. The ICC, however, consented to the violation because Greyhound claimed the transactions were too big—to undo them would be too costly.[38]

The company learns that chances of getting by with infractions are proportionate to their audacity, as Eugene Klein learned regarding Regulation T (see p. 97) and as International Telephone & Telegraph learned so well (see p. 278).

The Investment Company Act's rule that the number of bankers on an investment company's board must be held to a minimum shows that bankers' prominent role in investment and holding companies is nothing new. Even in 1913, decades before passage of the act, Louis Brandeis saw the danger to any corporation from interlocks with a bank (and to the bank as well):

> When a banker-director of a railroad decides as railroad man that it shall issue securities, and then sells them to himself as banker, fixing the price at which they are to be taken, there is necessarily grave danger that the interests of the railroad may suffer...The best bargains are made when buyer and seller are represented by different persons.

Brandeis saw that interlocking not only destroys loyalty. It softens judgment:

> A complete detachment of the banker from the corporation is necessary in order to secure for the railroad the benefit of the clearest financial judgment...It is outside financial advice which the railroad needs.[39]

Brandeis and the framers of the Investment Company Act would not have been surprised, then, that the ICC's own study of the Penn Central bankruptcy attributes largely to banker-directors the heinous ploy that the acquisitor pay dividends even while it incurred spiraling losses.[40] The banks represented on the board, according to the study, held great quantities of Penn Central stock in their departments and therefore voted for payments of dividends. As seen, banks represented by the acquisitor's directors, eager for the business from acquisition loans, laxly extended credit for financing the suicidal acquisition scheme. The Penn Central debt held by banks that interlocked with the Penn Central board far exceeded the cost of the companies acquired.

Public confusion as to which investments the securities statutes protect, and which not, is justified. The ICC classified as carriers corporations that are not carriers. Yet the SEC, which is responsible for enforcing the Investment Company Act, does not disclose which investment companies are free to operate outside of its enforcement. There simply is no list, said Chief Counsel Alan Rosenblat of the SEC's Division of Corporate Regulation:

> We do not have a list of publicly traded investment companies which operate outside the Investment Company Act since there is no legal requirement that they file anything with us.
>
> As a matter of law any company designated by the Interstate Commerce Committee [sic] to be a carrier is a carrier...[41]

No buyer of investment company stock should think the act protects him.

Better Than Transfusions of Public Money

Proof that legislators of the 1930s struck at the true causes of the economic devastation of their time is that circumvention of their safeguards gave rise, in the bankruptcy of the Penn Central, to exactly the type of corporate failure they sought to prevent. Rather than trying to prevent bankruptcies by transfusions of public money—a remedy the U. S. government would attempt some years later on a grand scale—a better solution would be to stanch that inexplicable circumvention of safeguards proven long ago.

II. Outside Operations

Federal securities law is only one paper guardian that should have flashed the Penn Central off Suicide Route. The other guardian seen in print but not action is government authority to prevent a railroad from taking over another corporation, whether it be in transportation or not. By Section 5 of the Interstate Commerce Act (54 Stat. 907), a railroad may not acquire another carrier without federal approval. Section 20(a) requires that a railroad obtain approval before issuing any of its "capital stock or bond or other evidence of interest in or indebtedness of the carrier." Congress enacted that section in 1920 in response to the ICC's own request for power to halt corporate acquisitions by railroads. The ICC had to close its eyes to its own reports of 1913 requesting that authority not to see the wreck that awaited its disregard of the authority—the wreck of Penn Central's takeover empire. The ICC had only to read:

> No student of the railroad problem can doubt that a most prolific source of financial disaster to railroads has been the ability of railroad managers to engage in enterprises outside the legitimate operation of their railroads, especially by the acquisition of other railroads and their securities. The evil results, first, to the investing public, and finally, to the general public, cannot be corrected after the transaction has taken place; it can be easily and effectively prohibited...

1. Every interstate railroad should be prohibited from spending money not in the operation of its railroad…
2. No interstate railroad should be permitted to lease or purchase any other railroad, without the federal approval…[42]

In March of 1969 the commissioners received another warning strikingly similar to the ignored reports of 1913, and again from their own staff:

> Management [of the acquiring corporation] may very well strip the carrier of additional assets reducing it to a corporate shell and then dispose of it.[43]

Northwest Industries gorged on, stripped, then threw off Chicago & Northwestern Railway. The acquisitor-holding company secured capital for its acquisition program from the railroad. After Northwest Industries had exhausted Chicago & Northwestern's tax benefits and credits, and paid dividends by selling the carrier's assets, it promptly began negotiations to sell the railroad. The acquisitor retains the name, but not Chicago & Northwestern Railway.

Benefits Like Magic

The government did not heed the warnings of its study of 1913 even *after* Penn Central collapsed. On August 15, 1972, ICC Commissioner Kenneth T. Tuggle informed the Sections of Public Utility Law and of Administrative Law of the American Bar Association:

> Today's Commission has not determined that diversification by conglomeration is necessarily bad for

transportation companies…Those lands and other rail assets can be released from certain government restraints, thereby facilitating their use in enterprises promising more attractive returns. The conglomerate, like magic, makes the tax benefits [from railroads' deficits] materialize.[44]

Nor has the collapse called attention to the commission's staff reports, primarily of 1969, that warned the ICC of the pitfalls of diversification by merger and the Penn Central's imminent bankruptcy. A special review of railroad acquisitors of March 11, 1969, warned:

Regardless of all the glowing self-serving statements made in current merger proceedings, we believe that these managements will involve the railroad in similar transactions [to that committed by Bangor Punta in sapping the capital of its subsidiary, the Bangor & Aroostook Railroad]. For certainly, holding company managements, in addition to their own strong self-interests, will owe their allegiance to the stock holders of the holding company and not to the railroad.

We believe the Commission should immediately exercise all of its available power to stop abuses before the assets are dissipated.[45]

Commissioner Tuggle presented the other side of the coin. Explaining to the assembled utilities and administrative law specialists, the "reasons why regulated carriers become affiliated" with acquisitors, he had just the remedy for old problems:

I believe it is conceded that diversification offers financial stability, especially to industries sensitive to economic fluctuations. The conglomerate can utilize the tax laws in dimensions not otherwise available; it can achieve economies of scale in management and administration; it can obtain credit on preferential terms and gain favor with investors; it has easier access to cash and capital.[46]

The commissioner did not document his findings. His staff's Merger Studies of March 26, 1969, refuted them all—except for tax gimmickry:

> The holding company [acquisitor] device…intensifies the Commission's regulatory and enforcement problems. The present threat to the public interest is…the dilution of the interest of managements of the conglomerates in the provision of efficient transportation service and the dilution of the capital structure of the railroads…Inclusion of railroads in conglomerate holding companies seriously impairs the Commission's ability to obtain information or to take timely action required for effective regulation of railroad rates and services.[47]

Commissioner Tuggle listed eight examples of grand benefits to railroads from being taken over—grand because the acquired railroads were not "in bankruptcy." Half of them, though, suffered "serious financial manipulations" by their parents, the Merger Studies revealed.

"With Southern Pacific, Illinois Central and others, the story is the same," Commissioner Tuggle said. "As [an independent] railroad, they enjoyed profits; but as subsidiaries in conglomerates, their profits increased."[48]

But of the Illinois Central and its parent holding company, Illinois Central Industries, the ICC staff studies said:

Illinois Central Industries charges Illinois Central Railroad in excess of the consolidated Federal income tax liability. Illinois Central Industries thereby took into income about $1.4 million more than its actual consolidated tax requirements.[49]

The Merger Studies found that Illinois Central Industries actually loaned Illinois Central Railroad cash for payments of dividends back to the holding company. "The practice of advancing funds with which to pay dividends," the staff explained, "is reminiscent of the abuses uncovered by the Federal Trade Commission in its investigation of the public utility holding companies during the 1930s..."[50]

Mr. Tuggle praised the schemes his commission found to be "not consistent with the public interest":

Greyhound presents a unique chapter. Until 1963 it was content to be a bus line, and in many respects it was No. 1. But then it formed a holding company and began to diversify. In 1969, it took over a giant conglomerate already in being. That was Armour-Dial. With annual sales of $2 billion, it was the second meat packer in the country...

Today it holds investments with a net worth listed at three quarters of a billion dollars, of which the bus lines make up less than one-third.[51]

But the commission had found that the Greyhound diversification-acquisition program so diminished the carrier's working capital that the former No. 1 hardly had money to keep its buses on the highways (see p. 236).

"Recognizing that the larger carriers are turning the conglomerate road," Mr. Tuggle added, "the ICC has proposed legislation to maintain the integrity of transportation regulation," i.e., to protect railroads from acquisitors that strip them of assets. But why would the ICC propose legislation to give it authority it already possessed? Clearly for one reason: to claim it does not have the authority.

Retirement Home

The ICC disclaimed much of its authority:

The 11-member Commission says that transportation reform is not its business. "Our authority," says ICC Chairman George M. Stafford [former attorney for the American Trucking Association], "is basically economic regulation." In other words, to regulate rates and conditions of shipping.

The authors of the ignored staff warnings against railroad diversification—and ever more concentrated control—did not agree:

The ICC staff, many of them long-time professionals, hardly bother to hide their scorn for the commissioners.

...Said another staff member: "The commissioners are plain wrong in saying that we have only rate-making powers. We have the power to require the railroads to do anything— make improvements, change routes, the whole thing."

...The Commission itself is considered to be a kind of retirement home for old politicians. A commissioner is appointed for seven years...The job is not demanding. The temptation to use it to reward old allies is too much for most Presidents."[52]

One appointee, Rodolfo Montejano, wondered why he was appointed:

I really don't know what all this is about. I was called by my friend Maramuto from the White House, who asked if I was interested. My background? The law firm I used to work for was the referral firm for the California Trucking Association. That's about all.[53]

American railroads derive wealth from land granted them from the public domain. The grants were an inducement for the original railway construction and its maintenance. Continuing his presentation to the American Bar Association, Commissioner Tuggle—a former lieutenant governor of Kentucky and a member of the commission for over twenty years—extolled the "vast acreage, producing income from minerals, lumber, agriculture and other non-rail activities."[54]

Investment of that national wealth in railroad facilities for the national good, as always intended, says the ICC Bureau of Economics, "yields a substantially high rate of return."[55] Nevertheless, Commissioner Tuggle must have slammed his fist on the lectern, so insistent were he and his colleagues that the wealth not remain devoted to public transportation.

He and his commission deduced a better use: financing of corporate takeovers. The acquiring headquarters will decide for itself how much railroad wealth the railroad needs and keep the rest for itself.

"The important question," is not how much an acquisitor takes out of a carrier, but whether the amount it judges adequate for transportation "is left in."[56]

Why Pay Taxes?

That property "left in" scarcely pays for taxes. The Association of American Railroads, in recommendations for "a more balanced transportation system," advises that "Railroads must be relieved of the burden of local property taxation":

> *The Federal government should exempt rail transportation from state and local taxation and reimburse the states for the revenue loss.*
>
> *Permissible types of state taxes should not, as a matter of Federal law, discriminate against railroads* [original italics].
>
> The railroads' financial weakness demonstrates the burden on interstate commerce which local property taxation of some $300,000,000 a year thrusts on the nation's rail arteries.[57]

The association figured out that not enough revenue is "left in" to pay for maintenance of rights-of-way. Its recommendation: "The Congress should require that the states devote 10 percent of Federal highway trust funds to grade crossing projects."

If the railroads could retain their birthright—wealth granted them from the public domain—they could, as intended, shoulder their state and local obligations the same as other taxpayers. The holding companies, however, applaud their Association of American Railroad's plan for continued transference to them of wealth from those public grants. After the acquisitors have taken over railroad assets, according to that plan, the government might just as well grant more public wealth to accomplish the purpose of the original grants.

Love for locomotives does not draw holding companies to railroads. Rather, railroads' wealth accumulated over centuries from public land grants is the real draw, the commissioner-erstwhile lieutenant governor said. He put his finger on why the acquisitor always (as the commission staff studies documented) decides the question of how much it "takes out" in favor of the nonrail operations.

According to available data, performance of acquired subsidiaries deteriorates more often than not even under an acquiring management that seeks to strengthen the subsidiaries' operations rather than transfer and deplete their assets (see p. 128). A subsidiary's future is bleak indeed, then, under such former managers as Heineman of Northwest Industries, Salgo of Bangor Punta, and Saunders and Bevan of Penn Central, who attached more value to railroad assets than to the railroad. Acquisitors must impress shareholders with earnings, or at least obscure declining earnings. Railroad assets are at their disposal.

"The Surface Transportation Board—which took over the remnants of the old Interstate Commerce Commission—yesterday approved the merger of the Union Pacific and Southern Pacific railroads into North America's largest railroad," *The Washington Post,* July 4, 1996, reported. "The merger has been a dream for a century. E. W. Harriman, the railroad baron who built the UP, bought 38 % of SP stock in 1901 and claimed control. But after a twelve-year battle with trustbusters of the day, the Supreme Court ordered him to sell it. The board's unanimous decision is almost certain to encourage another round of megamergers." Thus, the trustbusters' major turn-of-the-century accomplishment—dissolution of J. P. Morgan's infamous merger of Union Pacific with Southern Pacific—is now itself dissolved.

CHAPTER 12

The Securities and Exchange Commission

Corporations have caused the stock of other companies to rise merely by appearing to acquire them (see Table 2). The method for creating the appearance and the method for profiting are one and the same. The supposed acquirer's purchases of large blocks of the supposed target company's stock creates a rumor of an attempt to acquire. The price of the purchased stock then rises, especially when the purchaser issues a press announcement of intent to acquire. The purchaser profits from the sale of those blocks of stock, which it includes in its report of regular earnings. The report of increased earnings creates the false impression that the purchaser's operations are more profitable. That impression increases the market value of the purchaser's own stock. The increased paper wealth gives the purchaser greater resources to acquire more companies.

Gaseous Earnings

Gulf & Western Industries reported operating earnings of $72 million for fiscal year 1969. In fact, $32 million of that amount came not from operations but from sale of fortunate stock market investments—investments in enterprises that, reportedly, before sale, Gulf & Western was about to acquire.

On the morning of May 14, 1970, Securities and Exchange Commission Chairman Hamer Budge, General Counsel Philip Loomis, and Chief Accountant Andrew Barr appeared in the House

Judiciary Committee hearing room to explain why the commission allowed that deceptive practice of inflating operating earnings with stock market gains. The investing public had been led to believe, with the SEC's sanction, that Gulf & Western's entire reported earnings of $72 million were earnings from regular and continuous operations. At the time of the earnings report, the public had bid Gulf & Western common stock up to $53 per share. By the time of the hearings, when the public was becoming aware that the entire $72 million earnings of 1969 were not regular and recurring after all (1970 earnings declined to $44.7 million), Gulf & Western stock had fallen to $15.

The Securities and Exchange Commission exists to inform investors of the facts. The counsel asked SEC Chairman Budge whether Gulf & Western's addition of nonoperating earnings to operating earnings was lawful.

"I don't have a copy of your letter with me." The chairman could not respond.[1]

"We just looked into it a little," said SEC Chief Counsel Loomis. As to whether a noninvestment company could properly report stock market gains as regular profit, he said, "There was no formal examination to determine whether there was a violation of law. This was not pressed."[2]

SEC Chief Accountant Andrew Barr, however, met the issue head on. Gulf & Western's $32 million gain from the sale of securities was properly treated as regular, not extraordinary, income because, "I think you will find that that stock market gain has gone on for some time." The chief accountant was mistaken. During the previous fiscal year, 1968, Gulf & Western's gain from market investments was $4 million instead of $32 million. And before that, none at all.[3]

The committee turned to income reporting still more ingenious. How could LTV report that all companies under its control produced an aggregate profit of $2.3 million in 1969 when that alleged profit included tax benefits from an overall *loss* for that year? Included were $21 million refundable in federal taxes after a "carryback" of 1969 losses to previous years, and $19 million from predicted future tax savings, again from overall losses of the very year for which the profit was claimed.[4]

How could investors know from that reporting LTV subsidiaries incurred a loss of $80 million instead of the $2.3 million profit that LTV claimed? How could the public know the extent of the loss?

"I will see what I can do," the chief accountant replied. "I can't analyze it from here."[5] Nor could Mr. Barr analyze it back at his office with all the information the files of the Securities and Exchange Commission might offer.

The inside cover of the SEC handbook admonishes:

WARNING TO INVESTORS!
INVESTIGATE BEFORE YOU INVEST

Avoid unnecessary losses in the purchase of securities by following this ten point guide to safer investments: Before buying, Think! …

If you don't understand all the written information—
Consult a person who does![6]

If the friend the investor consulted in trying to comprehend the LTV income statement had been the SEC itself, he would have been wasting his time. Two weeks after the hearing Mr. Barr tersely replied by letter that a corporation's reporting of a $2.3 million profit in place of an $80 million loss is beyond the realm of the commission:

"Our files do not have sufficient information to answer this question."[7] What information the SEC files contain he did not say.

The commission did have available, however, a letter from LTV showing that the company had obtained SEC approval for reporting the $19 million of its 1969 loss as profit for that same year (in claiming future tax benefits from that loss).[8] The letter, a plea for the commission's leniency in the performance of its public trust, proves the effectiveness of unrefuted arguments behind closed government office doors, or closed automobile doors (see p. 282).

The letter and meaning of the law in question could not have been clearer. So clear that Mr. Barr himself said, "It is a factual matter." Bulletin II, paragraph 45 of the Accounting Principles Board, agreed upon by all parties as the governing regulation, provides: "tax benefits of loss carry-forwards should not be recognized until they are actually realized, except in unusual circumstances when realization is assured beyond any reasonable doubt at the time the loss carry-forwards arise."[9] Thus, a tax benefit from an annual loss may be used to reduce that loss only if income of future years to which the tax benefit can be applied is "assured beyond any reasonable doubt." A company's expectation of profit from ordinary operations, according to established, unquestioned, absolutely unequivocal precedent, does not meet that test. Nevertheless, the SEC allowed LTV to "carry the loss forward" and ignore the accounting principle.

The commission could hardly have chosen a worse time to flout the principle. LTV proved its purpose. The acquisitor lost money on consolidated operations in 1970 and 1971 also. No profit existed to which the tax benefit could be applied. Since the tax benefit expires after five years, LTV had, then, only 1972, 1973, and 1974 to produce income that might authenticate the $19 million benefit already reported as profit.

Even random examination by the House Judiciary Committee showed that the SEC had permitted other companies besides LTV to use the loss carry-forward. How often had the commission, which according to its own literature had been created "to provide disclosure of important facts so investors may make realistic appraisal of the merits of the securities and then exercise an informed judgment in determining whether to purchase them," permitted the hocus-pocus of telling retirees and other investors that a multimillion-dollar loss is really a two-million-dollar profit?

The question drew no prompt reply. The agency assigned to enforce rules of disclosure does not apply them to itself.

First Mr. Barr replied orally to the Judiciary Committee, "This is an area in which we are often consulting, that is right...I would like to try [to inform you of the number of times the SEC has allowed inclusion of future tax benefits in earnings reports for the year in which the loss occurs]."

But two weeks later the written response was that the number was anybody's guess, that such information is not in the SEC index: "a search of the indexed cards on tax accounting questions and recollections of the staff disclosed few instances of this question being raised. [But] The index would not necessarily include every case or telephone call where this question was raised."[10]

The SEC acquiesces to accounting gimmickry not countenanced even by corporations' own accountants. The auditing firm actually retained by LTV, Ernst & Ernst, refused to approved the breach of the tax benefit principle.[11]

Invisible Debt

But the accounting profession, as well as the SEC, liberally permits imaginative accounting. For example, the books of International Telephone & Telegraph show:

1. Merely by changing Continental Baking Co.'s depreciation accounting (from accelerated to straight line) upon acquisition, ITT reported for the subsidiary an additional $1.4 million of 1968 earnings.[12]

2. Solely by the same type of accounting change, ITT reported for the acquired Sheraton Hotel Corp. (and hence for itself) an additional $3.5 million of 1968 earnings.[13]

3. By reclassifying certain of Rayonier Corp.'s forestry expenditures as capital investment rather than as expenses (which diminish reported income), ITT reported additional 1968 Rayonier earnings of $1.1 million. "Instructions from ITT World Headquarters Comptroller's Office" required the reclassification even though it "has a substantial adverse effect on our cash generation," wrote Rayonier's management in October 1968. "[T]he loss of cash generation from the accounting change," the memorandum concluded, "will make it increasingly impossible to generate funds internally to meet our [Rayonier's] needs."[14] To Rayonier's profit for the year of acquisition, total accounting changes added $2.27 million— over one third of the subsidiary's profit increase.[15]

4. ITT's income for 1968, the year in which it acquired the three companies, increased artificially, then, not only from the addition of their ordinary earnings to the acquisitor's earnings. It increased also in the amount of $7.2 million resulting merely from bookkeeping changes—changes that at least one acquired company resisted.[16]

5. ITT World Headquarters transferred Avis Rent a Car's $5.5 million bank debt to ITT Credit Corp.—a subsidiary

"off" the acquisitor's balance sheet. Avis managers wrote President Geneen in February 1966: "ITT Credit Corp. now appears to be the vehicle to absorb this shift of obligations…"[17] In August they elaborated: "This does nothing for Avis, but hopefully, is helpful to our parent company."[18] Thus, the debt disappeared from the public ITT balance sheet not because it was paid but because it was shifted to an acquired company financially strong prior to takeover.

The House Judiciary Committee staff reported: "None of these accounting changes [resulting in reported profit, and hence market value, increases] were disclosed in the notes to ITT's financial statements."[19] Neither the accounting profession nor the SEC itself required that the investing public know that bookkeeping legerdemain, rather than proclaimed management ability, produced the nonrecurring profit increases.

Perhaps the SEC officials themselves are not to be blamed for the performance they revealed. The commission is understaffed. In 1970 Chairman Budge told the House Appropriations Committee:

> The public customer expresses frustration that he is unable to get a simple acknowledgment to his inquiry [to the SEC]. Additional complaint processors are urgently needed.[20]

Mr. Budge described the commission's inability to conduct even routine investigations of institutional investors:

> We face two real administrative problems in the investment company area. First, the failure of the Commission to be able to conduct the inspections of the investment companies, primarily the mutual funds. Second, processing the filings for registration.[21]

Two years later, on February 23, 1972, Mr. Budge's successor, Chairman William J. Casey, requested funds for 370 new staff positions, then reduced the request to ninety-four positions. The reduction would worsen the agency's performance, he told the House Appropriations Committee.[22]

But Chairmen Budge and Casey did not speak as typically beleaguered administrators short on funds. The SEC uniquely earns most of its appropriation itself. In 1968 it collected $14.6 million in registration fees—82% of its appropriations. In 1969 it collected in fees 118% of its appropriations. It was more than self-supporting, giving more to the U.S. Treasury than it received. Cost of operations, therefore, should not curtail performance.[23]

Payments Unpaid

Also a letter of July 14, 1972, from Chairman Casey to the Joint Economic Committee indicated that the commission did not have the staff to determine whether its own rules were being violated.[24] The Joint Economic Committee had asked the commission on June 19 whether Litton Industries had included any part of its fifteen-volume claims "summation" in its earnings statement.

The question is even more fundamental than the issue of inclusion of stock market gains in earnings reports. If a corporation can add the amount of *any* claim to its income report, that report can indeed be open-ended. The Joint Economic Committee asked for clarification of the commission's standards regarding the reporting of claims as income.

The inquiry hit a raw nerve: days before, Acting Commander R. C. Gooding, Naval Ships Command, had informed Litton that its claims summation concerning the cancellation of the four assault ship orders "is completely unsupported and is hereby rejected" (see p. 191). Litton's "LHA Program Reproposal," which included the claims, the commander said, "is almost completely unresponsive to the [contract] obligation."[25] But the Joint Economic Committee's attempt to determine whether Litton's earnings statement did include any of its navy claims was of no avail. The SEC simply could not answer the question.

The chairman's letter leaves no doubt, however, that Litton could inflate its income statement as it wished. The matter is for the company to decide, Chairman Casey explained:

> As to whether the inclusion of anticipated claim settlements as accounts receivable from the Government would violate any SEC rules, this would depend on numerous factors, the principal ones being the validity of the claim. As you can recognize, this is basically an area involving the judgment of the company's management and its independent auditors.[26]

The chairman assured, however, that any assault ship claims figure in Litton's reported earnings was inconsequential. Confidently he said the claims were "only a relatively small amount (when compared to total current assets)."

When compared to assets? Litton had $1.976 billion of assets. Litton's entire earnings of 1971 were less than 3% of assets.

The evasiveness of the SEC chairman's response to the letter's precise questions leaves little doubt that a corporate acquirer may indeed inflate its public earnings report with imaginative claims figures. Those figures create the false impression of greater profits, increase the acquirer's market value, and thus add to its resources for acquiring more corporations.

The breadth of Litton's imagination only began to come to light in December 1972:

Of a claim against the navy for $73.8 million, Litton reported $22.8 million as a "current asset." In fact, the navy agreed to pay no more than $7 million of the $73.8 million claim.

Of a claim for $94.4 million, Litton reported $10 million as "accounts receivable." The navy agreed to pay not one cent of the claim.[27]

In 1971 the SEC staff had specifically designed rules to protect the public from such imaginative reporting. Those guidelines would have warned investors of the exact defense production failures that gave rise to the ingenious claims. Rules proposed were:

At any time a material cost overrun has been incurred, the aggregate gross amount of such overrun should be disclosed.

[A government contractor] must inform its stockholders of known problems in meeting specification or delivery schedules which could result in additional expense to the company.[28]

After a series of secret hearings on the proposed rules, the commissioners rejected them. The Pentagon along with its contractors secretly objected to the proposals for public enlightenment. Rather than requiring adequate disclosure, the commissioners thought best to "appeal to defense contractors to make prompt and accurate disclosure," as the Stigler Report thought best that the U. S. president appeal to the nation (by televised address) to be more competitive (see p. 36).[29]

The commissioners actually professed belief that their plea would sway defense contractors. The General Accounting Office does not agree. It warns (Report B-163058 of July 26, 1973): "Litton is not likely to disclose [destroyer cost overruns and delays] until some time in the future when the full program seems committed" by Congress.

On December 20, 1972, the Joint Economic Committee of Congress inquired of SEC Commissioner Philip A. Loomis and SEC Chief Accountant John C. Burton (who succeeded Andrew Barr a few months earlier) of the commission's reason for the rejection of the disclosure rules. Commissioner Loomis, general counsel at the time of the rejection, first denied and then admitted that he had been involved in the decision. He also acknowledged that the SEC "should do much more" to protect the public from the claims of defense contractors. The reasons for the rejection of the disclosure rules, however, he did not state. Chief Accountant Burton did say that a defense contractor might be injured by "too much disclosure."[30] He didn't say how.

Less taciturn at the Joint Economic Committee hearings was the navy's civilian director of procurement, Gordon W. Rule. The appointment of Litton's former president, Roy Ash, as head of the U.S. Office of Management and Budget was, Mr. Rule stated, a "mistake." The procurement director testified that a few months before Mr. Ash's appointment, Mr. Ash had promised to go "on to the White House" to override the navy's reluctance to honor Litton's half-billion-dollar shipbuilding claims.[31] The promise carried increasing weight, for on December 14, 1972, Mr. Ash said, as seen, that he did not intend to disassociate himself from the Office of Management and Budget's decisions affecting the navy.[32]

President Eisenhower Insulted

Mr. Rule produced minutes of a meeting of June 6, 1972, recorded by navy officials:

Mr. Ashe [sic] indicated that it appears that some in the Navy have a built-in sense of self-righteousness concerning Litton's performance and that the Navy would have to relax this view if Litton is expected to proceed with the contract. Mr. Ashe indicated that he intended to meet with Secretaries Sanders and Warner and then on to the White House to explain the problem.[33]

Along with that warning Mr. Ash urged, according to the same minutes, that the navy allocate one or two billion dollars "to help out the nation's shipyards."[34]

Procurement Director Rule said he "did not like to see the Navy pushed around" and that if Litton could not deliver the ships as scheduled, the contract "ought to be terminated for default." President Eisenhower must be "twitching in his grave" over Roy Ash's appointment to be federal budget director, Mr. Rule said. Of course he remembered the president's dire warning against the "military industrial complex."[35]

Only hours later, the chief of Naval Material, Admiral Isaac C. Kidd, who had defended the selection of Litton for the ill-fated shipbuilding program before the House Armed Services Committee, arrived at Mr. Rule's home with a form for him to sign terminating his navy employment and some letterhead stationery on which to write his letter of resignation.[36] Less than two years before, Mr. Rule had received the navy's highest honorary award for civilian service.

Admiral Kidd "had lost considerable confidence in the judgment of Mr. Rule" and requested his resignation because, asserted Jerry W. Friedman, deputy assistant secretary for Department of Defense public relations, his remarks concerning President Eisenhower were disrespectful. Those remarks before the Joint Economic Committee,

Mr. Friedman said, were "very nearly the ultimate in poor taste and bad judgment."[37]

The SEC's leniency extends not only to acquisitors who report as income their questionable claims against the government but also to those who consent to SEC suits for securities law violations. On Friday, June 16, 1972, the Securities and Exchange Commission filed suit against International Telephone & Telegraph Corp., ITT Senior Vice President-Counsel Howard J. Aibel, ITT Secretary-Counsel John J. Navin, and ITT's servicing banking firms—Lazard Freres & Co. and Mediobanca di Credito Finanziario of Italy. On the following Tuesday the commission disposed of the litigation by accepting the defendants' consent to permanent injunctions which only forbid their further violations of securities laws. Obviously, the injunctions require of the defendants nothing more than the law already requires of everyone.

The SEC alleged that ITT and its two officers had dealt illegally with ITT stock. ITT violated Sections 5(b) and 17(c) of the Securities Act of 1933, the complaint charged, by not disclosing in the registration of company stock its settlement with the Justice Department of the Hartford Fire Insurance Co. merger case (see chapter 14). That settlement required ITT's divestiture of major acquired subsidiaries on which ITT relied for a substantial portion of its earnings.

Illegal Profits Kept

Aibel and Navin violated Section 10(b) of the act, according to the charge, by the fraud of selling ITT stock while knowing as "insiders" of the undisclosed Hartford settlement. "Insiders" could accurately anticipate (as the suit for injunction claimed) that the requirement for divestiture, once public knowledge, would cause a sharp decline in the market price of ITT stock.

On June 17, 1971, the day after the settlement but before the public knew of it, Aibel sold his 2,664 ITT shares for $164,000. Navin sold 1,500 shares for approximately $100,000 on July 16.[38] On July 31, ITT publicly announced the settlement and the required divestiture of the profitable subsidiaries. On the first trading day thereafter, ITT stock fell by $7 per share.

ITT, Lazard Freres, and Mediobanca were further charged with violations of Section 5(a) and (c) of the act for unregistered transference of Hartford stock. The transaction accomplished the defendants' purpose of creating the semblance of a sale, which was sufficient reason for the Internal Revenue Service to free from taxation the acquisitor's purchase of Hartford stock (see chapter 13).

The SEC did not inform the Internal Revenue Service of the violations against which it secured the injunctions because, the commission explained, "it was not the SEC's responsibility to call to the attention of another Government agency an action that might or might not be a violation of laws enforced by the other agency."[39] For that novel doctrine commission officials cited no precedent.

The judgment to enjoin the defendants against further violations—consented to in a matter of hours—contained the strange provision that they did not admit those violations. That a person can agree to be enjoined against violations and not admit them is fiction. The fiction served to allow the defendant ITT officials to retain the profits (approximately $7 per share sold) secured from the defrauded stock purchasers who bought before the announcement on July 31, 1971, of the required divestiture. Hence, the judgment entailed no penalty.

Conversely, previous SEC suits against "inside" dealing have required the restoration of profits (to stock purchasers who incurred losses because they did not have the confidential information known by the parties with whom they traded). The SEC's policy had been to require restoration. The commission gave no explanation for abandoning that policy.[40]

Section.15(b)(5)(c) of the Securities Exchange Act and Section 9(a)(2) of the Investment Company Act prohibit enjoined parties from engaging as broker-dealers of securities. Granting further leniency, the SEC declined to enforce those provisions against ITT (owner of ITT Hamilton Management Corp., a manager of mutual funds, and of ITT Variable Annuity Insurance Co.) and Lazard Freres. The prohibition is automatic. Nevertheless, the SEC permitted the enjoined firms to continue to engage in the trading of securities.[41]

In reaction to that frustration of the law's intent, on September 21, 1972, the chairman of the House Committee on Interstate and Foreign Commerce, Harley O. Staggers, and the chairman of the Finance Subcommittee, John E. Moss, sought to examine the ITT documents on which the commission had based its suit for injunction. Congress created the SEC by delegating to it congressional power; therefore the commission is an agency of Congress. By requesting the documents, the House Commerce Committee exercised its responsibility to oversee the SEC's performance of Congress's mandate.

SEC Chairman William J. Casey acknowledged Congress's right to the documents. But on September 22 he refused the congressional staff access to them. Only the whole commission could grant the permission, Mr. Casey said. While the House Commerce Committee waited on that "whole" permission, the SEC without notice carried the ITT documents to the Justice Department—an executive agency beyond the reach of Congress.[42]

Mr. Casey, informing the House Commerce Committee on October 6, 1972, of the transfer, stated that the Justice Department out of the blue had requested the documents two days earlier. SEC

Deputy Director of Enforcement Stanley Sporkin, however, not only testified to the committee that Justice Department officials never asked for the documents prior to the congressional request for them, but that on September 21, the day of the committee's original request, those officials told the SEC that the Justice Department did not want the documents and not to send them over.[43]

The two chairmen could only respond, "The Securities & Exchange Commission, in clear disregard of its duty to cooperate with the House Commerce Committee, has undertaken a most injudicious course of conduct," and ask:

> By hindering the [Committee staff] from gaining access to these materials prior to the time when the Members must recess, the SEC perhaps expects it has won the day... Are the so-called "independent regulatory agencies" really independent of the Executive Branch of Government?[44]

In December of 1972, Mr. Casey announced plans to resign as SEC chairman, get away from Washington, and join the diplomatic corps. If the congressional oversight committee does pursue its investigation of the application of securities laws to acquisitors, it should hear the familiar reply (see p. 201): The responsible official "is the little man who...went away" yesterday.

CHAPTER 13

The Internal Revenue Service

Acquisitors offer the owners of target companies quick market gains. Special tax treatment prevents taxation of the gain. It is a government gift for fueling the thriving trend of corporate takeovers.

A taxpayer who sells his investment in a corporation to reinvest in another must pay tax on any gain from the sale. Naturally, that requirement restrains the taxpayer from changing the form of his investment: since he must pay tax on any gain from the sale, he will not sell in order to purchase shares of another company unless that company offers investment opportunities that outweigh the tax liability. The fact that he sells and purchases not to realize gain but only to place his funds in a more diversified company or one that offers more liquidity (i.e., is more widely traded) makes no difference—he must still pay tax on any gain from the sale.

Subsidy for Acquisitors

However, stockholders who transfer their investment to an acquisitor by an exchange of stock (not by selling and purchasing stock) even for sheer market gain are not required to pay tax on the gain at the time of the exchange. If shares selling for $10 on the market are exchanged one for one for the acquiring corporation's shares which are selling for $15, the tendering shareholders incur no tax liability from the exchange even though it increases the value of their investment by 50%. Thus, a restraint that applies to other forms of reinvestment does not apply to corporate acquisitions consummated by exchange of stock.

Naturally, tax specialists refer to the exemption for such acquisitions, or "reorganizations," as a subsidy for corporate concentration. The exemption is also known as a "tax avoidance leak," although the gain from the conversion of acquired-company stock into the acquisitor's shares is taxable upon sale of those shares. Former Internal Revenue Commissioner Sheldon S. Cohen, in the *American Bar Association Journal,* January 1969 (p. 40), quotes an authority, Randolph Paul, who wrote in 1940:

> The exemption of corporate reorganizations has been called a subsidy; the question has been raised whether the net effect of the provisions, particularly as they stand in the statute with all their particularity, is not unfortunate, in that in actual effect they are a serious tax avoidance leak and one of the major and indispensable forces in the thrust toward economic concentration...

The tax exemption for corporate acquisitions evolved after the First World War, when Congress, fearful that industrial dislocation would result from conversion to a peacetime economy, actually sought temporarily to encourage mergers that served a business purpose. But once created, the exemption has proven indestructible.

The House Ways and Means Committee recommended elimination of the provision in 1933 after a study of tax loopholes. However, as Mr. Cohen writes:

> [The Department of the] Treasury came to the rescue, arguing that because of the depression most reorganizations

[mergers or acquisitions] were being undertaken to reduce corporate capital structures...In addition, Treasury felt that substantial corporate adjustments would have to be made in recovering from the depression and that imposition of a tax on such "legitimate" reorganizations would be poor economic policy.

World War I and the Great Depression have come and gone; yet the exemption remains.

To qualify for the tax advantage, Section 368 of the Internal Revenue Code, the merger must amount to a "reorganization" which serves a "business purpose." Deputy Commissioner of Internal Revenue William H. Smith told the House Judiciary Committee that shareholders cannot use the provision as a scheme for tax avoidance:

> As the law on reorganizations took shape, important judicial doctrines and tests were also evolving. These included the "business purpose" doctrine....
>
> In order for a reorganization [i.e., merger or acquisition] to qualify as tax free, the business purpose doctrine requires a showing of a bona fide business purpose. This rules out reorganizations conceived for the avoidance of Federal income taxes.[1]

Successful requests to the IRS for the exemption often implore, however, that if the IRS does not grant it, the parties will abandon the merger plan. Thus, they reveal that the proposed merger is "conceived for the avoidance of Federal income taxes" on market gain. That the merger hinges completely on the granting of the exemption rules out any other motivation.

Managements, urging shareholders to approve merger, extol the exemption to them. They again reveal the purpose of merger to be market gain, a "business purpose" that, the deputy commissioner assured, unequivocally "rules out" granting of the exemption. Yet the IRS invariably grants the exemption.

For example, the management of Landis Tool Co. informed its shareholders by a proxy statement of November 25, 1967, in which it solicited their approval of a merger with Litton Industries, that the first of two purposes of the "reorganization" was to double the market value of their investment by securing Litton stock, which sold for more than "twice the price" of Landis stock. The second purported purpose was the standard unsubstantiated praise for diversification: "Landis production is concentrated in grinding machines...Litton manufactures hundreds of different products."

In successfully seeking a ruling from the Internal Revenue Service of nonrecognition of gain under Section 368 by asserting that the merger served a business purpose, the Landis management made no mention of that prime purpose of market gain. Nor was it mentioned to the Federal Trade Commission, which heard from Landis that the reason for the merger was only for shareholders to acquire an "equity interest in a larger, more diversified corporation, the shares of which are to be listed on the New York Stock Exchange."[2]

Eyes Closed to Reasons

To find a "business purpose" of a merger—and accelerate the pace of industrial concentration and create a tax revenue loss estimated by the Federal Trade Commission to exceed a billion dollars annually in 1968 and 1969—the government must, then, close its eyes to the reasons given to the stockholders to induce their approval. The IRS's nonrecognition of gain requires its nonrecognition of the actual reasons for corporate acquisitions that are contrary to the rationale of the Section 368 exemption.

Other transactions of Litton Industries are cases in point. Had the IRS collected taxes on the market gain (to shareholders of acquired companies) from exchanges of stock vital to the acquisitor's growth and essential for its reporting of steadily increasing acquired income, Litton probably could not have acquired a hundred corporations.

The assets of Litton Industries rose from $119 million in 1960 to $1.580 billion in 1969.[3] The method of acquisition followed the classic (now familiar) cycle. The parent company inflated its earnings reports with the income of the companies purchased with Litton preferred, or preference, shares. Because the acquisitor determined its earnings per share by dividing total earnings only by the number of common (not newly issued preferred) shares, reported earnings per share steadily rose. Credulous investors thereby believed that Litton income would continue to rise even without mergers. They bid steadily higher for Litton shares, giving the acquirer more paper wealth with which to buy more corporations.

Litton common stock rose from 25.5 in the spring of 1964 to 114 in the fall of 1967. Reported income rose from $29.8 million in 1964 to $70 million in 1967.[4] Not disclosed was that the purported rise in earnings came from dumping the profits of the acquired companies onto the pile and not from the subsidiaries' heightened performance after merger.

Displaying that universal acquisitor characteristic, Litton does not release earnings statements of the individual acquired subsidiaries. For that information, showing deterioration of performance after acquisition, would dispel the myth of the parent's management

ability, which is essential to acquiring the income of more companies, in turn essential to inflating more earnings reports, in turn essential to driving Litton stock prices still higher.

The excuse for keeping subsidiaries' performance secret is that
competitors somehow can use the information. There was no fear
of the Securities and Exchange Commission's questioning that excuse and disrupting the cycle. Hamer Budge, chairman of the commission while that cycle functioned most feverishly, believed that
to require disclosure that so much as showed investors whether an
acquisitor purchases earnings (by acquiring other corporations) or
actually profits from operations might be too "complex." He explained to the House Judiciary Committee on May 14, 1970, that
investors had

> asked us to go to the point that if you are manufactur-
> ing cookies, they would want a breakdown as to the sales
> and profits on vanilla cookies and on chocolate cookies.
> Obviously, that is no real importance to the investor in
> making his determination as to whether or not he wants to
> buy or sell the stock.[5]

In the eyes of the commission, subsidiary operations as diverse
as machine tools, textbooks, myriad electronic equipment, food,
submarines, typewriters, and cash registers (only to begin the list of
Litton's acquired product lines) were no more worthy of distinction
than baking processes for different flavors of pastry.

The committee's investigation disclosed that of the fourteen
post-1964 acquisitions which had assets of $10 million or more
each, Litton could show only one (Jefferson Electric Co., which
operated at a loss before merger) that yielded more earnings from
its assets under Litton management than before merger.

While that fact was unknown, the myth that Litton management improves the performance of acquired subsidiaries easily persisted. Just as easily, stockholders of target companies agreed to exchange their ownership rights for ever-appreciating Litton shares.

So rapid was the appreciation, the market value of Litton stock some shareholders received doubled the value of the stock they gave in exchange. In all cases, the gain to the acquired shareholders, the inducement for the trade, was substantial.

The facile generalizations concerning diversification and management ability cited to target-company shareholders to induce them to expect that market gain to continue and to agree to merger are essentially the same as those which the acquisitors cite to the IRS to explain "business purpose," and thus obtain a favorable tax ruling. The *Conglomerate Investigation Report* of the staff of the House Judiciary Committee of June 1, 1972 (p. 362), describes that technique as "overstatement":

> Litton's image-making has developed flamboyant sham into an art. Overstatement is a way of life. It led Litton in the hearings to assert that even its organization chart was so special that it was imbued with business confidentiality. In the process of developing its image, Litton has utilized all of the sophisticated accounting techniques and statistical gimmicks available. It is adept at concealment, misdirection and incomplete statement.

On September 21, 1964, Litton informed the IRS that its acquisition of Hewitt-Robbins, a materials handler, was for "important business reasons." Hardly more explicitly, the letter continued, that upon merger, Litton would have greater opportunity to utilize "its electronic know how and equipment."[6] The Hewitt-Robbins stockholders would benefit from being able to "participate in a combined company having a substantially greater amount of assets, capitalization, earnings and sales volume and a greater degree of diversification." The IRS was not told that the Hewitt-Robbins shareholders consented to the merger only after being assured of a gain of 25%—paid in Litton stock—over the pre-merger market price of the target company stock.

The "business purpose" of the Rust Engineering Co. takeover, as explained to the government by letter of April 4, 1967, was that "Litton sees the opportunity to provide a real service to the people" by entering the heavy engineering-construction field in the United States and abroad—especially in underdeveloped countries.[7] The combination, it said, would create a capability for "rebuilding of cities and construction of totally new cities, combating water and air pollution, harnessing of atomic energy and utilization of space and undersea exploration." Neither the opportunity nor the capability has been heard of since.

Further examination of explanations of merger (for which the IRS granted the nonrecognition of gain ruling) from to-be-acquired managements to their stockholders casts doubt that the motivating purpose of "reorganization" was any other than the chance to secure the more highly valued Litton stock.

For example, the management of New Britain Machine Company, while urging its shareholders to approve merger with Litton, asserted in its proxy statement of November 7, 1968:

1) Litton's stock will pay a higher dividend than the New Britain common shares [even though New Britain earned more than Litton];

2) New Britain owners will receive in exchange stock listed on the New York Stock Exchange;

3) The transaction will be tax free;
4) Employee benefits will be continued;
5) New Britain, as a subsidiary of Litton, will have substantially greater financial and technical resources for product development and other purposes;
6) Greater diversification will result from merger;
7) Value of stock received: $42.50 market value of New Britain share for $65 market value of Litton share.

Obviously, only one of those assertions could be a motive for merger. If the New Britain managers wanted to give shareholders a higher dividend, they had only to declare it from their earnings, greater than Litton's.

The second assertion—that New Britain stock would be converted into securities listed on the New York Stock Exchange—is only slightly more absurd as an independent reason for corporate "reorganization" than the fourth assertion—that employee benefits would be continued.

The fifth assertion—that New Britain after merger would have greater resources for research and development—is highly suspect as a purpose of merger. Litton announced no New Britain product innovations after acquisition that were not possible before. Also, Litton management replied orally in answer to congressional inquiries that product development in its machine tool division was carried on not in separate research facilities but rather on the production line.

The sixth explanation—that merger would result in diversification—is said of any takeover by a giant acquisitor.

One explanation stands alone as a convincing reason for the shareholders. Overnight, as a result of the merger, the market value of their investment would increase more than 50%—from $42.50

to $65 per share. And because the IRS was able to discern a "business purpose," it would obligingly free the gain from any present tax.

The opportunity takeovers afford for quick market gain renders the "nonrecognition of gain—corporate reorganization" ruling a greater force in the "thrust toward economic concentration" now more than ever. With that ruling, combining shareholders secure their objective of financial gain and avoidance of tax—an objective for which, the deputy commissioner told the House Judiciary Committee, the IRS does not grant the ruling.

Cash Purchase Concealed

The IRS further contends that it applies the Section 368 "tax free" merger ruling only to legitimate corporate "reorganizations." A "reorganization" merger, according to the IRS's regulations, can result only from the exchange of voting stock between the owners of the two combining corporations—not from the acquisitor's cash purchase of the target company's stock.[8]

ITT Vice-President Charles T. Ireland thus informed ITT President Harold S. Geneen on January 2, 1969, that International Telephone & Telegraph Corp.'s cash purchase of 8% (1.74 million shares) of Hartford Fire Insurance Co.'s outstanding common stock would jeopardize their chances of securing tax exemption for ITT's acquisition of Hartford.[9] For each of the remaining outstanding common Hartford shares, valued at $38.25, ITT would offer one ITT preferred share, valued at $49.[10] The 28% market gain was the inducement to Hartford shareholders to accept the trade and consent to takeover. If the IRS did not grant the exemption and thus required the tendering Hartford owners to pay tax on that gain in 1969, Vice President Ireland feared that the stockholders would probably reject the trade. Nevertheless, he recommended to

President Geneen that ITT continue, before the exchange offer, to acquire as much Hartford stock as possible with cash. The IRS could be dealt with later.

His advice was sound. The gamble succeeded. In spite of the cash purchase of stock the IRS recognized ITT's Hartford acquisition as a "reorganization" and granted the Section 368 tax exemption. The granting was lawful because ITT sold its cash-purchased Hartford Stock, the IRS said. The transaction that the IRS identified as a sale was ITT's temporary transfer of the Hartford stock to an Italian bank, Mediobanca of Milan. In fact, the shares passed from ITT to Mediobanca—and after the ruling was secured back again from Mediobanca to ITT—without even the alleged purchaser's payment of a purchase price.[11]

Other elements of the transfer which indicate that Mediobanca acted as ITT's agent for the temporary removal of the shares rather than as purchaser are that:

> ITT paid Mediobanca an agent's fee of $1.3 million for undertaking the transfer.
>
> ITT and not Mediobanca, by their agreement, would incur all gain or loss upon any change of value of the Hartford stock after its transfer to Mediobanca.[12]

In spite of those elements inimical to a sales transaction, the IRS asserted with ITT that the transfer constituted an "unconditional bona-fide sale to an unrelated third party."[13] ITT's winning of the tax break for its Hartford Fire Insurance Co. acquisition raises the question: When does the IRS *not* grant the Section 368 "subsidy" for corporate concentration?

Table 6

Tax Treatment Accorded 411 Large
Acquisitions° in Mining and Manufacturing, 1963–1968

	Number of tax-free exchanges	Percent
1963	25	71
1964	36	84
1965	37	88
1966	47	96
1967	101	86
1968	104	83
Total	350	85

° "Large" acquisitions are defined as those involving acquired firms with assets of $10 million
or more.
SOURCE: *Economic Report on Conglomerate
Mergers,* p. 145.

There's no way to know. The IRS declined to turn over any
written opinion setting forth its reasoning for bestowing the exemption. The IRS replied only: "We believe that members of the
American public justifiably have come to expect the Service to ob-
serve this policy of confidentiality and that it would be improper
for us to deviate from it."[14]

Of eighteen acquisitions in 1967 and 1968 of corporations
with assets in excess of $250 million, fourteen were "nontaxable."
Further, the number of "tax free" rulings increased with the inten-
sity of the merger movement (see Table 6).

CHAPTER 14

The Department of Justice

The tale never ends, Antitrust Chief McLaren must have mused as he heard the regulators recite their failures—without even knowing of Bank of America's future takeover of a thousand banks and receipt of the $43 billion bailout. He resolved more strongly that International Telephone & Telegraph's takeover of Hartford Fire Insurance Co. must not proceed, that regulation by competition and the "invisible hand" must survive.

He was right. The regulators' tale did not end. "Eric Kolchinsky is one of a series of whistle blowers who tried to tip the Securities & Exchange Commission to wrongdoing, only to be ignored. This failure helped continue two of the most celebrated frauds of the last decade, costing unwitting investors millions of dollars," *The Washington Post*, January 21, 2010, reported. The celebrated frauds involved dangerous approvals by Moody's, the credit rating firm, of viral subprime mortgages, and could not have happened without the serial bank acquisitions Mr. McLaren for a time thwarted merely by announcing intent to fight the Hartford takeover. One acquisitor that received bailout billions, JPMorgan Chase (see p. 4) provided hundreds of millions of dollars for the illegal trades, the *Post* continued.

1

The Penn Central managers sought government salvation when their corporate acquisitions led to their company's ruin. They failed and accepted bankruptcy. The managers of International Telephone & Telegraph asserted that their acquisition course was equally as ruinous, sought government rescue, and succeeded.

The cost of the International Telephone & Telegraph rescue was higher than that proposed for Penn Central. The cost was not just abandonment of litigation to protect "the invisible hand" of market structure by restraining giant mergers between two corporations that have separate markets, a goal the Attorney General and his experts proudly announced to Congress in 1970. The cost also was the reversal of antitrust policy: the demoralization that permits the giant combinations of corporations even though their markets are the same—even though they are direct competitors (see chapter 3).

The success of ITT's plea for help and the failure of Penn Central's plea show that selecting the right governmental bargaining technique is more important than the cause itself.

The decision to acquire Hartford Insurance Co. required ITT's ability to win over government officialdom. Relying on that talent alone, ITT laid its very existence on the line. Failure to surmount the government's obstacles to the merger, ITT declared after purchase, would have ruined ITT. Those lethal barriers not only loomed but at times prevailed over the acquisitor. But along the harrowing course ITT never once admitted defeat, not even official defeat.

Attorney General John N. Mitchell branded the attempt to acquire the second largest fire insurance company as illegal before ITT even announced the plan. The Justice Department would sue to prevent the merger, he said, between any of the largest corporations and any other occupying a dominant position in its industry.

Besides confronting that obstacle, ITT had to win the consent of the Connecticut Insurance Commission. The commission's disapproval—at one point final—could be no less a disaster than the federal prosecution.

A Nation's Penalty

No matador ever walked into a ring confronting such overwhelming odds as did ITT after paying a $500 million premium for Hartford shares. Few matadors have risen and won after initial failure to fend direct impact of double onslaught.

ITT prevailed where mighty challengers have failed because it convinced the government the company had reached the ultimate height of syndicate growth: size is so great, that penalty to the syndicate is penalty to the nation. Because of vastness of acquired operations, the Justice Department reasoned, to prosecute ITT for its illegal actions would injure the entire economy. The acquisitor successfully argued: better to allow a violation of law and public interest, and not prosecute, for what harms ITT harms the USA.

ITT secured special dispensation by asserting, however contradictorily, that a first grant of leniency had harmed the corporation. The government was therefore obligated, it successfully pleaded, to grant further and greater leniency.

In 1968 the serial acquisitor took over at the rate of one corporation every two weeks. In 1969 the pace more than doubled to an acquisition every six or seven days. Shortly before the start of that year, ITT purchased 6% of the outstanding stock of Hartford Fire Insurance Co. On December 23, 1968, the acquisitor proposed merger—the largest of all time. The Hartford management agreed on April 10, 1969.

The insurer was attractive to ITT the same way Great American was to National General, and Reliance was to Leasco. Hartford possessed, as the Netter Report disclosed, great capital resources (see p. 92). Its investments in securities of other companies had substantially appreciated over the years. Income from them, paid as dividends and interest, had more than doubled between 1958 and 1968 when ITT began purchasing Hartford stock.

But far more important, the sale of the securities would result in an overnight profit exceeding $200 million. Accounting rules would allow an acquirer to add any amount of gain from the sale of those securities to the income statement of any year it chose. Thus, ITT would obtain not just capital assets from the merger but nearly a quarter-billion dollars to report as increased profitability, i.e., as proof of World Headquarters' vaunted management ability for which investors paid dearly.

On August 1, 1969, the Department of Justice applied for an injunction against the merger, asserting "it may substantially lessen competition…in violation of Section 7 of the Clayton Act." But the U. S. District Court for Connecticut denied request for injunction on October 21. ITT had been reporting increasing profits, its stock was at new heights, and Hartford shareholders voted overwhelmingly to approve the merger in June 1970. They gave ITT 99.8% of the insurer's outstanding common stock in exchange for ITT preferred stock. ITT then controlled Hartford. The district court, however, ordered the acquisitor to maintain Hartford's business separately from its own while the government's suit was pending, in case the court should order that the acquisition be undone. The federal prosecution was only the second obstacle ITT was to confront. First there was the state barrier.

Insurance is Connecticut's basic industry. For its protection, the legislature enacted Public Act 444 that obligates would-be acquirers of insurance companies to disclose extensive information about themselves revealing whether the purchaser's control will be in the insurer's and the public's interest. Efficiency of the acquiring management is, then, a major issue: data showing the management's performance are most pertinent.

At hearings before Insurance Commissioner William R. Cotter on November 14, 1969, however, ITT gave no information about the revenue and income of its individual acquired subsidiaries, which would indicate how well ITT managed them.[1] It did not supply the data, later obtained by the House Judiciary Subcommittee, which compare post-merger profitability and management efficiency of ITT-acquired companies with their premerger performance. Those data indicate that operations of the companies deteriorate under the acquisitor's control (see p. 128).

Nor did the acquisitor disclose that the preferred shares it proposed to issue in exchange would pay upon conversion (to common shares) smaller dividends than the common shares which the Hartford owners would surrender.[2] Also undisclosed was that once ITT possessed the Hartford shares, it would greatly increase the dividends and pay them to itself, adhering to established pattern of insurance company acquisitors (see chapter 6).

After holding public hearings, the commissioner ruled on December 13, 1969, that the proposed acquisition would be inimical to Hartford's as well as the public's interest. He officially forbad the merger.

In appreciation, the Hartford City Council passed a resolution in praise of Commissioner Cotter and his forthright decision. As the decision *then* stood, the corporate headquarters of Hartford Fire Insurance Co. would not pass from its city of birth.

Hearing in the Limousine

Six months later, on May 23, 1970, Commissioner Cotter changed his mind. During those months ITT officials often met with him. Of course behind closed doors. Far from ever having the opportunity to rebut ITT's second round of arguments, representatives of the public never even knew of them.

During the time of those *ex parte* encounters, Commissioner Cotter became a candidate for Congress. ITT proposed that its previously acquired Sheraton subsidiary build the Sheraton Hartford. The hotel would contribute greatly to Hartford's economy, they assured. ITT would even help construct Hartford's new convention center.

One day in May, Commissioner Cotter rode with ITT officials in the ITT automobile to examine possible sites for the Hartford Sheraton.[3] The next day he approved the merger. Previous accumulations add to the growth of a rolling snowball.

The commissioner based his consent to the merger on the very evidence he had based his refusal. No new evidence or substantial change of acquisition plans prompted his change of mind.

ITT did not rest on the laurels of victory. A professedly more determined foe had already struck. As expected, the Department of Justice had lost in the district courts its attempts to prevent ITT's purchase of Grinnell Corp. and Canteen Corp. Those suits, also alleging violations of Section 7 of the Clayton Act, were companion litigation to the Hartford suit. Assistant Attorney General McLaren, not at all surprised by the lower court decisions, eagerly prepared for appeal to the Supreme Court.[4]

Over and over again the Department of Justice had expressed unequivocally its resolution to secure judicial interpretation of antitrust legislation that would restrain the trend of corporate concentration. Its confidence was well founded. Since the enactment of the Celler-Kefauver Amendment twenty years before, the government had never lost a merger suit appealed to the Supreme Court.

ITT even admitted its despair of winning the Hartford suit in the Supreme Court. But as seen, the acquisitor was not to swerve from its chosen course. Thus impended, by all appearances, a test of implacable wills.

As with the Insurance Commission encounter, however, the struggle was not to occur in a room of open and free debate, but behind closed government-office doors. The outcome, too, was the same. ITT won.

For two years Justice Department officials had been telling Congress they would secure the ultimate judicial test of whether the laws Congress had already enacted were sufficient to cope with the most prolific trend of industrial consolidation. They expressed complete confidence that their arguments to the courts would prove the laws adequate. Their words, then, negated any sense of urgency that Congress enact stronger antitrust statutes.

The Justice Department's complete about-face—its acquiescence to ITT's takeover of Hartford—without even requiring ITT to reduce acquired assets, should hardly convince acquisitors that they need fear enforcement of the Clayton Act and the Celler-Kefauver Amendment. The Justice Department claimed, nevertheless, that the settlement restrained ITT from further growth by merger. ITT's acquisition rate, true indeed, was down from a corporation a week. During the first nine months after the Hartford settlement, however, the acquisitor easily acquired sixteen more companies, six of them in the United States. They included a producer of steel tubing, a manufacturer of packaging tools and supplies, a manufacturer of boiler controls, a bakery, a life insurance company, an exporter of hams, a pump production operation, a producer of toiletries and drugs, and a brassware firm.[5] A few days after the announcement of the settlement, *Business Week* of September 7, 1971 (p. 84), reported that the Justice Department had thrown away its chance to "define the limits of corporate growth in a modern society." It lamented:

> Antitrust chief Richard McLaren himself acknowledges that he missed a shot at a landmark decision because "it is a lawyer's job to get a satisfactory conclusion to a case as rapidly as possible."
>
> [This is not the right attitude] for the chief architect of the government's antitrust policy.
>
> If McLaren and his lawyers feel that the mushroom growth of the conglomerate corporations threatens the U. S. economy and infringes the antitrust laws, then it is their duty to push the point with the courts until the law is established beyond question. It is not fair either to the companies involved or to the public to keep on brandishing the antitrust gun without ever proving that it really is loaded.

Assistant Attorney General McLaren resigned his office in January 1972 and became a U.S. district judge in Chicago soon after. In March, when the Senate Judiciary Committee asked him why the Justice Department threw in the sponge, Judge McLaren explained that, besides his own "background of twenty-five years," three factors predicated the decision not to seek divestiture of Hartford.[6]

The first was the opinion of ITT counsel Lawrence Walsh—a former U. S. district court judge and former deputy attorney general. At the time of the settlement negotiations, he was chairman of the American Bar Association's Standing Committee on the Federal Judiciary, which passes on qualifications of candidates whom the Justice Department recommends for federal judgeships. The second factor was the opinions of other government agencies, namely, the Departments of Treasury and Commerce, and the President's Council of Economic Advisors. The third was the report of a certain Richard Ramsden, a New York investment advisor.

On April 16, 1971, Lawrence Walsh urged Deputy Attorney General Richard Kleindienst to drop the department's suit, two years in the preparation, for Hartford divestiture. "To us this is not a question of the conduct of litigation in the narrow sense," he said in a memorandum to Kleindienst:

> Looking back at the results of government antitrust cases in the Supreme Court, one must realize that if the government urges an expanded interpretation of the vague language of the Clayton Act there is a high probability that it will succeed.[7]

The ABA federal judiciary chairman/ITT lawyer argued to the Justice Department that the issue of ITT industrial consolidation was too important to be determined by the laws of the United States. An issue of such "economic consequence" was too far reaching to be entrusted to the Supreme Court's interpretation of those laws: "the Supreme Court…will disregard any economic or other public benefits resulting from the merger," he informed Kleindienst.[8]

Further, the Justice Department learned, "It is our understanding that the Secretary of the Treasury, the Secretary of Commerce, and the Chairman of the President's Council of Economic Advisors all have some views, with respect to the question under consideration."[9] Therefore, those gentlemen—to whom ITT representatives had *ex parte* access as they had to Commissioner Cotter and did not have to Supreme Court justices—should decide the question of divestiture, Walsh concluded.

Once again the specter of the Stigler Report appeared. Two and a half years earlier, its authors had recommended to the president "retrenchment" of antitrust enforcement. Quoting from that very report, Walsh's memorandum urged that all opposition to takeovers cease while the government undertook still another "comprehensive"— i.e., protracted— investigation of mergers.[10] Whether the findings of that proposed study would serve any better purpose than the Federal Communications Commission investigation findings now kept under lock and key, or its investigation findings the Federal Trade Commission fed into the paper shredder—as were documents from ITT files explaining government negotiations leading to the Hartford settlement—the Walsh memorandum did not say.[11]

Use Sellers as Buyers

Though the Justice Department had been investigating and preparing its Hartford case for over two years, Walsh felt no awe in submitting time-worn generalizations that the department's own evidence, set forth in its Grinnell trial brief, refuted. With no countervailing evidence, his memorandum recited the litany:

A ban on significant mergers and diversification would injure vital national interests.

Diversification by merger is the most important guarantee that every economic resource of the Nation will in fact be used to the best advantage.

Through diversification, scarce management skills, additional resources of capital and know-how, and most important, the will and ability to plan growth, can be brought to bear in new industries.

A Supreme Court decision…would have exactly the same immediate impact as a statute and would be even more difficult to modify as experience showed its unwisdom.[12]

"Thus diversification by merger is often the only effective means of stimulating new competitiveness in established industries. ITT is a case in point," Walsh informed the deputy attorney general. But if he had read his department's own trial brief Mr. Kleindienst would have seen absolute refutation. It cited specific examples of ITT's securing of reciprocal sales that, far from "stimulating," circumvent competitiveness. The brief (Civil Action 13319, United States District Court, District of Connecticut) stated:

ITT has also engaged or attempted to engage in reciprocal practices over the years. Although ITT has issued an "anti-reciprocity policy" directive, this directive...does not forbid the use of reciprocal practices when convenient and useful to ITT. In fact, Mr. Harold Geneen, supposedly the strongest advocate of ITT's alleged "anti-reciprocity" policy, and Mr. Howard Aibel, ITT's General Counsel...have themselves resorted to the use of reciprocity.

The brief spoke of the sale of electronics equipment:

According to the testimony of Mr. Aibel, ITT had been trying to sell some of its equipment to General Telephone & Electronics Corporation (GT&E) with little success. However, when a GT&E subsidiary inquired about the possibility of selling some of its services to ITT in Puerto Rico, Mr. Aibel and Mr. Geneen seized upon this inquiry as a "golden opportunity" to suggest to GT&E that ITT would consider GT&E's offer and, in return, GT&E should consider ITT's offer to sell equipment to GT&E. GT&E correctly interpreted the offer from Mr. Geneen as an offer to enter into a reciprocal arrangement with ITT. Internal documents from GT&E's files reveal that GT&E considered various reciprocal arrangements with ITT, all of which involved the allocation of GT&E business to various ITT subsidiaries.

Avis Rent a Car was an ITT triumph of market rivalry, in Walsh's eyes. "The improved competitiveness of the companies ITT has acquired is illustrated by the growth and development of subsidiaries like Avis and Sheraton," it assured the Justice Department. "ITT in other words, has been able to apply modern management skills in such a way as to increase very substantially the efficiency and competitiveness of the companies acquired."[13] The memorandum did not draw from the House Antitrust Subcommittee investigation, which disclosed that Avis's performance steadily improved before ITT took it over and steadily declined afterward, in spite of envisioned anticompetitive advantages.

For proof directly opposite to Walsh's claims, the Justice Department officials again had only to read their own brief:

> Maximizing of sales through use of ITT's purchasing power and influence was also a consideration in the ITT-Avis merger...Various ITT subsidiaries have engaged in reciprocal practices. However, the only reprimand for such activities by ITT management resulted when the Government's discovery in this lawsuit uncovered reciprocity within ITT. In fact, Mr. Aibel testified that the only way ITT would ever uncover reciprocity within its confines would occur if the perpetrator of the act brought it to the attention of ITT management.

Abrasive as the Walsh presentation may have been to the staff that prepared that brief, its chief received it with equanimity—

and worse—conviction. "Mr. Walsh and I are very close friends and have developed a very close friendship over the three years as a result of our work together in the judicial program [for selection of federal judges]," Deputy Attorney General Kleindienst testified.[14]

Judge McLaren, while head of antitrust enforcement, however, was not convinced:

> For example, he [Walsh] said in there, as I recall, that our policy was stopping perfectly normal, legitimate mergers that had nothing to do with effects on competition, and I strenuously argue with that...
>
> I say again, I strongly objected and was not persuaded as to the legal aspects of it [Walsh's argument].[15]

The only factors, then, that could have caused his change of mind (according to his testimony) were the opinions of the three government agencies and of the financial analyst Richard Ramsden.

But the heads of the three agencies, Judge McLaren explained, were in favor of the suit to force ITT to divest Hartford. In contradiction of Walsh's information to the deputy attorney general, Judge McLaren testified:

> I thought that Dr. McCracken [Chief of Economic Advisors]...was very much in favor of our antitrust policy, and I have never heard...that [Commerce] Secretary Stans or the Treasury people were against it, and I subsequently turned out to be right.[16]

The former antitrust chief appeared to be ill at ease during the hearing. He did not live long after assuming his judgeship.

The report of the certain Mr. Ramsden, then, remains the predominate, if not sole, factor for the decision to allow ITT to keep Hartford. The Justice Department retained Richard Ramsden to evaluate the merits of the Walsh memorandum. Although the department had its own staff of economic advisors, already well versed on the subject of ITT finances, the officials thought best that an "outside" analyst judge whether ITT should takeover Hartford.

Open-Ended Catastrophe

A prominent conclusion of the Ramsden Report and the memorandum it evaluates was that divestiture of Hartford would hinder ITT's ability to contribute to the U. S. balance of payments. The Walsh memorandum did not even claim that the Hartford takeover would bring back dollars from abroad. Ramsden asserted, however, that takeover of Hartford would be so beneficial to ITT that Europeans would rush to buy more ITT stock.[17] By that reasoning, no multinational corporation should be denied any anticompetitive advantage. President Geneen extolled "bring back" before the Senate Judiciary Committee on March 15, 1972:

> Any company such as ITT which is scheduled during the next ten years to bring back something on the order of $5 billion from abroad without imports to the credit side of the United States balance of payments cannot be said to be working against the national interest in these trying times.[18]

Ramsden also discovered "fundamental changes in the insurance industry suggesting... earnings improvement" because insurers are assuming less risk. Less risk, yes. Hence, impaired public service precisely because of the phenomenon he praised—acquisitors' takeover of insurance company assets. He had the figures:

> The amount of insurance capacity in certain markets has been reduced. One estimate is that $1.5 billion of capital has been taken out of fire and casualty companies (mostly in the form of dividends paid to holding companies).[19]

Consequently, insurance companies have "radically reduce[d] their exposure to unprofitable markets such as high risk urban areas"—areas in dire need of it. All too clearly, Ramsden told the Justice Department, "excessive competition has been reduced." Did he know that department's purpose was to *protect* competition?

But most important would be the effect of the takeover on ITT's ability to report increasing earnings and hence continue to acquire:

> Most likely it [prevention of the merger] would result in further concern as to ITT's ability to manage consistent earnings increases and such concern would probably be reflected in a diminished multiple on the common stock.[20]

That concern, Ramsden wrote on May 17, 1971, would likely result in a decline of ITT's stock value from $64 to $54, or 16%. The paper loss to the acquisitor's investors would be approximately $1.2 billion. His reasoning persuaded the Justice Department to accept the Walsh recommendations. Such a decline resulting from a court-imposed divestiture would be so harmful to the U. S. economy and ITT's investors, the department agreed, that its antitrust policy must be halted and the suit abandoned. In return, ITT consented to acquire no more U. S. corporations with assets exceeding $100 million. Also, the acquisitor agreed to divest itself of four subsidiaries. By selling them, however, ITT would only exchange those assets for other assets, such as cash. Thus, there would be no decrease of ITT's total assets as a result of the divestiture order.

No precedent exists for the government's reversing its course to prevent a 16% fluctuation in a stock's quoted value. And fortunately not, for in April 1972, ITT's market performance proved the futility of the government's compassion. At the very time that the Senate Judiciary Committee hearings were examining the events leading to the abandonment of the suit, the acquisitor's stock fell to the *exact* forecasted $54 level! The stock thus incurred the 16% decline in spite of the allowance of the combination that was supposed to prevent it.

Surprisingly, the decline did not result in the economic collapse the Walsh and Ramsden memorandums had vividly forecast a year earlier. ITT director Felix Rohatyn described that previously envisioned catastrophe to the committee in April 1972 with irrefutable reasoning—irrefutable because so vaporous:

> It had been our belief that a forced divestiture of Hartford Fire could raise fundamental issues of national policy transcending both the narrow scope of traditional antitrust philosophy and the narrow interest of ITT. Several points were involved, including possible United States balance of payments effects and government policy toward preserving the underlying strength of domestic companies competing overseas. In addition, as chairman of the New York Stock Exchange Surveillance Committee during that period of financial crisis, I was concerned that so massive a divestiture might unsettle our securities markets, and with possible impact on some financial organizations.[21]

Mr. McLaren actually claimed that he took that reasoning (as later expressed) at face value. He informed Deputy Attorney General Kleindienst on July 17, 1971:

> We have had a study made by financial experts and they substantially confirm ITT's claims as to the effects of a divestiture order. Such being the case, I gather that we must also anticipate that the impact upon ITT would have a ripple effect in the stock market and in the economy.

"Such being the case," he acquiesced to dropping the lawsuit to prevent the acquisition:

> I say reluctantly because ITT's management consummated the Hartford acquisition knowing it violated our antitrust policy; knowing we intended to sue; and in effect representing to the court that he need not issue a preliminary injunction because ITT would hold Hartford separate and thus minimize any divestiture problem if violation were found.[22]

Mr. McLaren meant: The court permitted ITT to acquire Hartford even while the Justice Department sued to undo the acquisition—but only with ITT's promise that it would hold Hartford apart from its syndicate while the court heard the suit. If the court ruled against acquisition, it did not want ITT to be able to say: "We cannot divest Hartford now because it is an integral part of our syndicate." But ITT did *not* hold Hartford apart as it had promised. Thus, it successfully made the exact argument the court carefully sought to avoid: undoing the acquisition would result in injury to the ITT syndicate, to Hartford, and to the nation.

The consented final judgment which permitted ITT to retain Hartford contained only two restrictions: it enjoined ITT for ten years from acquiring more U. S. corporations with assets over $100 million and restrained it from forming reciprocal sales agreements with its suppliers. The government thus closed its eyes to the greatest anticompetitive advantage ITT intended to secure for itself and Hartford. That oft-described benefit, carefully explained in other ITT merger documents, is the use of the nation's third largest mass of corporate employees as captive customers of their employer's goods and services.

Self-Patronage Generator

At the very time Justice Department officials were acting on Walsh's denials of anticompetitive intent, they had right in front of them the ITT merger plan of November 2, 1968, entitled "Tobacco-ITT Joint Opportunities." Tobacco was the acquisitor's code name for Hartford Fire Insurance Co. The unpublished document reads:

MARKETING OPPORTUNITIES WITHIN THE ITT SYSTEM

1. ITT is a vast consumer of insurance products. Most of our 300,000 [400,000 by 1972] employees in the United States and overseas are covered by various types of group insurance. In addition, the corporation has a substantial volume of property and casualty insurance in the United States and overseas.

2. ITT has 150,000 employees in the United States with a similar number overseas. This total places ITT about fourth [by 1971, third] in the lists of largest employers. We have developed a combination life

> insurance/mutual fund program for sale to those
> ITT employees in the United States through pay-
> roll deduction. Virtually all of these people require
> various forms of casualty insurance which could be
> readily included in the salary savings program...

Again, the merger plan was to increase the acquired company's profitability not from improved service, but from its syndicate
membership that generated captive clientele—an anticompetitive
luxury for which other insurance companies would pay with their
loss of customers.

Besides subsidiaries' employees, subsidiaries' customers would
tie in as Hartford's "captive audience." The document continued:

6. There are several opportunities for the marketing of
   insurance programs to special ITT interest groups:

   a) Sheraton has 1.2 million credit card holders.
   b) Avis has 1.5 million credit card holders.
   c) 100 million APCOA [an ITT subsidiary] parking
      transactions.
   d) ITT has over 200,000 shareholders.

It is suggested that various types of insurance programs
could be offered that may or may not prove advantageous
(Avis cars for salesmen, Sheraton hotel arrangements, etc.,
TDI, travel advertising locations, etc.)

7. Levitt has sold homes to 80,000 homeowners and
   building new homes at the rate of 6,500 per year.
   Within five years this figure will exceed 11,000. These

purchasers require homeowners and mortgage insurance which may be offered through special marketing
programs.

And do not forget the business of financing syndicate operations, the document concluded. It virtually described a self-
contained economic domain, a planned economy, an insulated
generator of self-patronage, a cocoon.

At the end of 1971, ITT allayed the Justice Department's fear
for the acquisitor's ability to report consistent earnings increases.
Its income for 1971 was $44 million more than the 1970 figure. The
12% increase was right within the area of ITT's traditional objective.
The department, then, should be quite satisfied with its acquiescence
to the combination. For $36 million of that $44 million rise is the
amount that ITT secured as the overnight profit from the sale of
Hartford investment securities.[23] Undoubtedly, the sale of the insurance company assets produced that sizable earnings increase which
ITT attributed to management ability. As seen, the Hartford acquisition supplied the acquirer with almost a quarter billion dollars of
asset appreciation that can be used for ready profit in future years.

In reply to the question in 1970 of whether ITT planned to
keep on acquiring until Doom's Day, President Geneen told the
House Antitrust Subcommittee:

No; I think our fundamental objective [has been accom-
plished]...

Chairman CELLER. When do you think the saturation
point will be reached?

Mr. GENEEN. That is a good question. In my opinion,
the saturation point is reached when we feel that we can't

manage things that we would be acquiring, and I think we are getting to a point—[24]

But can an acquirer dependent on acquired earnings to maintain a history of increasing earnings ever reach a "saturation point"? At the May 10, 1972, ITT shareholders' meeting, Mr. Geneen urged "transformation" of U. S. antitrust laws to allow more mergers so as to prevent "weakening of the nation's economic strength." Quoting Secretary of the Treasury John B. Connally, he called for: "[t]urning antitrust policy inside out, so that the government would *encourage* mergers instead of *discouraging* them [original italics]."[25]

The acquisitor's power, as Napoleon said of his own, depends always on expansion (see p. 7). As long as the acquisitor, like the emperor, can convince the right people that its expansion and the nation's destiny are one, the cycle should continue unimpeded.

2

Constantinople never recovered after its cavalry charged into the maintain-pass at the battle of Manzikert on August 26, 1071. The Turks dealt the deadly blow nine hundred years to the day prior to the Justice Department's abandonment of the ITT-Hartford suit. But the imperial city somehow survived in its death throes for centuries amid lesser disasters—even amid seeming triumphs.

A lesser disaster to the open market system, heaving in its death throes for decades after the scrapping of the ITT-Hartford suit, was the combination of Time, Inc. with Warner Communications, with Turner Broadcasting Co., with AOL. Time purchased Warner in 1990 for $14.9 billion in cash and stock. After purchasing Turner Broadcasting in 1996, Time Warner, Inc. combined with AOL in December 2000 at a cost of $164 billion. Although approved by the Federal Trade Commission, the amalgamation reported a loss of $99 billion for 2002—then the greatest corporate loss of all time. In December 2009 it admitted failure and wrote off the colossal merger expense by separating from AOL. If the Justice Department had persevered against ITT's takeover of Hartford and won judgment in the Supreme Court, as even ITT itself expected, it would have secured precedent lethal to the Time-Warner-Turner Broadcasting-AOL mergers.

The merger of Chrysler Corp. with Daimler AK also inflicted a lesser disaster to the open market guided by the "invisible hand." Daimler paid $36 billion for Chrysler in 1998 but sold it to Cerberus Capital Management in 2007 for *less than zero*. Daimler paid the equity fund $725 million to take the automaker off its hands. As *Daimler Company News*, May 14, 2007, grimly told stockholders: "Daimler will transfer...Chrysler completely *free of debt*. Net cash outflow resulting from the transaction will be EUR 0.5 billion [$725 million]." The ITT-Hartford suit thrown to the winds August 26, 1971, would have militated against that merger also.

And like Constantinople, the open market claimed seeming triumphs in its twilight era. American General Life Insurance Companies "completed the largest acquisition in the history of the life insurance industry in 1982 with its acquisition of National Life," its annual report of 2002 boasts. The report extols at least three more insurance-company takeovers by the Texas insurer, each flowing over the billion-dollar mark—each inspiring "consolidation in the industry." The last sentence tersely explains "consolidation": the colossal acquisitor is now itself a subsidiary of an acquisitor that dwarfs it—American International Group (AIG) with "distribution channels in 130 countries and jurisdictions throughout the world." The seeming triumph catapulted AIG toward the trillion-dollar mark.

Once over the mark, it collapsed. Of course taxpayers became liable then to AIG for $180 billion. AIG acquired Bulgarian Telecommunications Co., valued at over $2 billion, only months before the collapse, when Pierre Mellinger, head of AIG Capital Partners "tried to appease the markets about AIG's limited exposure to the subprime market and its healthy cash flow," *Forbes*, August 17, 2007, reported.

During those last months AIG dared not slacken. In the field of power and water operations alone it grabbed MTC Holdings, among America's largest terminal operators; U. S. port operations from Dubai Ports World; and Utilities, Inc., a holding company controlling small water utilities spread across 17 states. It also took over the UK lender, Ocean Finance & Mortgages, Ltd.

Other acquisitors, as well, use mergers to obscure imminent doom. Two months after the Federal Reserve Board unanimously approved the application of Cleveland's National City Bank (among the nation's largest subprime lenders) to acquire a Chicago bank, the acquirer reported a $19 million loss for third quarter 2007. It never reported a profit again. PNC Financial Services of Pittsburgh in turn acquired National City's assets. Likewise, the fourth largest U.S. bank, Wachovia, purchased the California mortgage-lender, Golden West, in 2006. In 2008 the Federal Reserve Board ordered Wachovia to cease operations. Wells Fargo, now the fourth largest U.S. bank, acquired its assets, *The Washington Post,* December 21, 2009, reported. Hence, even acquisition failures augment asset agglomeration.

American International Group's takeover of the insurer SunAmerica Inc. in 1998 through an $18 billion exchange-of-stock, of course treated as pooling of interest (see chapter 13), was almost as vital to its size and complexity as its purchase of American General. "That complexity means its tentacles are spread throughout the financial system so that its collapse could be catastrophic,"

The Wall Street Journal, September 18, 2008, reported. The domino effect "could reach around the world."

AIG lent $2 billion of the $180 billion collected from taxpayers to its acquired subsidiary International Lease Finance Corp., *The Los Angeles Times,* October 20, 2009, reported. ILF Corp. buys Boeing and Airbus aircraft and leases them to seventeen airlines. It's the world's largest aircraft lessor. ILF Corp. needed the billions to pay debts to Citigroup, itself a beneficiary of bailout billions (see p. 4). AIG had already lent ILF Corp. $1.7 billion earlier that year.

Do all firms such as ILF Corp. in need of $3.7 billion to survive secure it from taxpayers? *The Los Angeles Times* didn't say. It had said enough. Firms not large enough by themselves to win bailout billions may still win by combining into an acquisitor whose tentacles grasp the billions for them.

The tentacles need not belong to AIG. Bank of America has never said it needed $43 billion from taxpayers before it took over a thousand banks after August 26, 1971—some directly, others by taking over their acquisitors. It does not refute that it funnels billions to those banks. Nor does it contend the banks could have secured the billions on their own—or even needed the billions before being trapped in the acquisitor's subprime mortgage meltdown.

Two disasters share the same August 26th conception date. Whether the meltdown and the Turks' ravishing of Constantinople share the same renown for infamy, let time tell.

Icebergs Dead Ahead

By turn of the nineteenth century, a schoolchild could see that great corporate amalgamations set prices and killed off competitors. Thus, Congress designed antitrust law to prevent formation of monopolies—to protect "the invisible hand" of competitive market structure. By mid-twentieth century that law prevented enterprises operating in the same industry from combining.

Blocked in the traditional channel of homogenous combinations, the forces of industrial concentration overflowed into the channel of heterogeneous combinations. Combinations of corporations functioning in dissimilar industries—merged by acquisitors who brought them under "operating umbrellas"—soon surpassed the merger trends of decades past. Guardians of market structure studied the new trend and saw that giant mergers between dissimilar enterprises could be as harmful as combinations of competitors—and of suppliers with their customers.

They brought suit to prevent the heterogeneous combinations and to establish that unbridled combinations of corporate giants are dangerous even if they do not create monopolies. The guardians saw court victory within their grasp. But suddenly they walked off the field.

The very auspiciousness of their litigation was a kiss-of-death to antitrust enforcement. Never again could the public take antitrust enforcement seriously, authorities wrote. Time has borne them out:

now acquisitors are practically as free to combine competing corporations as they are to combine non-competitors functioning in dissimilar fields.

The acquisitors are right in making no distinction between competitors and non-competitors. For the evidence that the guardians carefully prepared for asserting the public's rights under the Sherman Antitrust Act but then threw away shows that undue heterogeneous combinations no less than combinations in like industries hide the same age-old icebergs:

1. The increased size of acquiring corporations from their takeover of other companies' does not increase productivity. It does not yield resources for product development. Bigness and innovation don't mix, Antitrust Chief McLaren said:

> The bulk of the available evidence runs counter to the hypothesis that high concentration, huge size, and substantial market power are prerequisites for research and innovation. Indeed, some of the most careful studies find that if anything, market power and the security of bigness, with the concomitant vested interest in the status quo, may have a stultifying effect. And I submit that this should not be surprising.[1]

The agglomerators, nevertheless, expound the theory of "economy of scale." Data of acquired companies' performance after merger, reluctantly revealed, contradict any idea that acquired scale yields economy. The data indicate that proficiency of acquired companies declines more often than not under their new agglomerated management.

Acquisitors claim to provide financial resources to companies they take over. Data show just the opposite: they acquire to transfer those companies' financial resources to themselves. Thus, capacity for improved performance—such as reduction of cost to consumers—generally diminishes after a company loses its independence.

2. The increased control and mass of operations of acquiring corporations, while not resulting in more efficient production, do provide profitable opportunities for circumventing competition. The securing of anticompetitive advantages is the one contribution that acquiring managements may bestow on acquired subsidiaries.

From sheer vastness of acquired assets, an acquisitor will arrange to increase the sales of acquired companies by delivering to them a clientele composed of its acquired subsidiaries, its employees, companies beholden to it for its patronage, and even of those companies' employees. If the acquisitor has acquired to the extent that it is the third largest employer in the nation, the captive clientele may be a greater sales factor than the quality of its products and services.

3. Acquisitors obscure management deficiencies through production for the U. S. government. Defense contracts, easily secured outside the confines of competitive bidding, may be performed at a standard unacceptable for commercial production.

Acquiring headquarters claim to possess management skill even in diverse industries foreign to them. Post-merger performance data of their diverse takeovers disprove that claim, as seen. But because the government is not an exacting customer, a corporate acquirer that fails the test of commercial production may expect to profit from shoddy production performed for a government agency. The Defense Department, rarely critical of a private contractor's failure to perform as contracted, fails to enforce even contract provisions that would insure that the public receives what it pays for.

4. The "hidden asset value" that acquisitors discern in target companies most often is the chance for reporting increased profits

303

through simple bookkeeping alterations. By merely changing the acquired company's conservative methods of accounting for inventory and depreciation, the acquiring corporation reports overnight multimillion-dollar paper profit increases. The paper reports are the basis of the acquisitor's vaunted claims of superior management ability. Those assertions founded on accounting gimmickry are the traditional basis for a rise in the public market value of the acquiring corporation. That rise stimulates the acquisition spiral. With the artificially increasing value of its market shares, the acquisitor has greater wealth with which to continue to acquire.

5. Acquisitors report increased profits and claim management proficiency also by adding earnings of acquired companies to their own earnings figures. By purchasing those companies with preferred stock or debt securities rather than with common stock, acquisitors avoid having to issue new common shares. Thus, they artificially report greater earnings spread over the same number of shares.

After takeover the acquisitor may shift the purchase debt (represented by debt securities) to the target company. That debt requires higher prices. No one pays it but consumers.

6. Lending banks and borrowing acquisitors form virtual acquisition trusts through interlocking of officers. In return for the favor of acquisition loans, the acquisitor may transfer to the bank the banking business of the corporations it purchases with the bank loans. Federal law once prohibited commercial banks from using their depositors' money to gain control of other corporations. But that law had to be repealed to erase the illegality of a giant commercial bank's takeover of a giant investment house (see p. 4). That bank and an acquisitor secure near-infallible methods for stock market manipulation (see chapter 4).

7. Traders of securities and acquisitors also form takeover trusts through interlocking of officers. Officers of a brokerage

firm who are directors of an acquirer persuade it to purchase a corporation whose stock the brokers already own. When the acquirer begins to purchase that stock (at a premium above the market price) the brokers can sell at a higher price. Goldman Sachs co-president, Jon Winkelried, explained in *The New York Times*, December 16, 2009:

> One goal was to have Goldman wear several hats in deals, for example, by advising on a merger, financing it, and investing in the transaction.

The purchase is profitable to the brokers also because they serve as the agents for both the purchaser and the seller. They receive commissions for shares bought (by the acquisitor) and for shares sold (by the target company's shareholders).

Brokers instigate corporate consolidations by informing failing or debt-saddled corporations of companies which, over the years, have built up great amounts of cash or easily liquidated reserves. After merger those acquisition targets, having lost their capital resources to acquisitors, are unable to provide their former quality of service to the public. They've thrown their good money after bad.

8. Acquisitors' anticompetitive practices boomerang. Acquisitors then try to shift their losses to the taxpayers. Roy Ash, former president of Litton—acquisitor of Ingalls Shipbuilding Corp.—urged that the U. S. undertake a multi-billion billion-dollar program of direct aid to failing shipbuilders, while he was U. S. budget director. Penn Central won administration support for its plan to transfer acquisition losses to the taxpayers. LTV-Memcor salvaged itself, after undercutting competitors' bids, by shifting its losses to the Pentagon.

Charles Bluhdorn, CEO of Gulf & Western, urged in the company's "Report for Nine Months Ended April 30, 1973" passage of "…legislation giving small investors up to $5,000 of tax-free earnings from gains in stock market securities, to attract the millions of individual investors back to the nation's capital markets" after a speculation bubble burst. The bubble arose in the first place largely from the predictable, confidential market fluctuations he engineered with Chase Manhattan Bank before it took over J.P. Morgan Co. in 2000 and bailout funds in 2009. The recommended billion-dollar plan of tax-free earnings for small investors would yield a rising market and gain to acquisitors from any future stock manipulation.

9. The very industries that agencies are to regulate come to dominate them. Assistant Attorney General of the Antitrust Division Thomas E. Kauper explained on October 27, 1972:

> The industry to be regulated may at first view the regulation as inhibiting, but over the years learns to control the regulation to its own purposes either by its expertise in dealing with regulation or by becoming the primary "constituent" of the regulators.[2]

Merger advocates unabashedly describe to government officials an insurer's weakened ability to serve the public after its takeover, but even then request a suspension of antitrust enforcement that would allow the greatest corporate consolidation of all time. And the officials grant the request.

Government regulatory agencies permit mergers that their staffs urgently advise against. Contrary to the laws they exist to enforce, the agencies permit the acquisitor to report inaccurate and misleading performance data. The public is denied, then, its right to know the true financial condition of companies in which it invests. Concealing that knowledge from investors enables an acquirer to acquire until it is bankrupt.

By claiming jurisdiction over an industry but failing to assert it, a regulatory agency negates the authority of other regulatory agencies over the industry. Investors consequently lose protection that Congress fully believed it had provided.

Pleas to government agencies to permit corporate consolidation by ignoring antitrust laws—and indirectly by ignoring investor protection statutes—stridently portray emergencies. Secretary of the Treasury John B. Connally warned that if antitrust enforcement is not curtailed so as to permit more mergers and more anticompetitive practices, the nation may face outright "revolution."[3]

Threats to open market structure, however, never appear to be emergencies. Action to protect it can always be postponed and then forgotten—until the bailouts and anticompetitive meltdowns turn catastrophic.

The public's safeguards against corporate consolidation are suspended after officials hear behind closed doors uncontested pleas that enforcement of the safeguards will cost the acquisitor its $500 million gamble that the safeguards will not be enforced. Pleas citing the acquisitor's loss of stock premiums paid, the supposed loss of the shareholders' market value, and the supposed contribution to U. S. balance of payments—however spurious—are always clothed in concrete monetary terms. The granting of those urgent pleas is only at the cost of an abstract principle—free competition ordered by the "invisible hand."

☆ ☆ ☆

As the number of autonomous industrial units diminishes, so declines the power of the people to regulate their economy naturally by choosing among market rivals. As control of industry falls into fewer and fewer hands, so increases viral contagion of schemes for fast money (credit default swaps, for one)—and the opportunity of industrial suzerainties to enjoy reciprocal patronage. Through heightened concentration, industrial assets are regrouped not for the needs of the public but for the advantage of the concentrators.

Acquisitors profess to altruistic use of their concentrated power as though their benevolence were fair exchange for the public's loss of authority. But their economic power easily converts into political power.

ITT, "Serving People and Nations Everywhere," was spending $1 million in Chile in 1970, for "low cost housing and other ventures" similar to "Marshall Plan aid to help the Chilean economy and reaffirm that the company has confidence in Chile," it told the Senate Foreign Relations Committee on March 22, 1973."[4] The committee revealed ITT documents of September 1970 showing the true purpose to be, rather, "to induce [through concert with other international corporate control centers] economic collapse in Chile" in order to prevent the Chilean government from exercising its lawful authority over ITT operations.[5]

The public's decline of control over the economy envelops even the corporate managers themselves. Litton Industries insisted that the House Judiciary Committee maintain the secrecy of Litton's organization chart. The committee repeatedly inquired why the chart most commonly found in corporate reports should be a secret. Finally, the reply came that its secrecy was to prevent the company's own managers from viewing the chart. The sight of the vastness of the scheme of acquired operations, the executive committee chairman explained, would demoralize Litton executives:

Mr. McDANIEL. [Litton executives] are proud, jealous of their prerogatives…By looking at the box where they go on the organization chart, they become dissatisfied with their status in the organization and it affects their morale. This is the reason we do not like to see the organization chart made public.

There are men who run organizations that are not shown on the chart, who pride themselves on having direct access to Messrs. Thornton and Ash...They look at this chart, and they see two or three layers in between and it affects their morale.[6]

If loss of independence demoralizes even the corporate managers, whom but the acquisitors does it inspire?

Iceberg Alert

The acquisitors' letters Mr. McLaren read while preparing suit to block then history's largest merger alerted him to icebergs that *change shape but never melt* (see p. 4):

- *Audacity of takeover plots.* The grandeur itself of mega takeovers stultifies shareholders' good sense.
- *Schemes made to appear indecipherable.* Machinations for buyer to be seller and seller, buyer (to name one), must be simple to work. Easiest questions penetrate the camouflage of Ponzi schemes.
- *Assurances that times have changed.* Proven safeguards of market structure have become millstones (such as separation of depository and investment banks), acquisitors claim. They target the most successful safeguards.
- *Divided authority.* When companies are stacked on top of another, who is in charge? Who is to blame for investing in Madoff securities (see chapter 1)? Even organization charts obscure chains of command.
- *Viral contamination.* Loss of autonomy renders a thousand purchased banks defenseless against the disastrous contagion of their acquirer's subprime mortgage risks (see p. 2).

Corporate tentacles would not extend through society (see p. 299) had Mr. McLaren's Justice Department heeded those warnings and, victory in hand, stayed the course.

"Bernanke urges financial regulation to prevent crises," *The Washington Post* by-line, January 4, 2010, read. But the Federal Reserve chairman did not urge protection of the market's free "invisible hand," such as separation of commercial and investment banking. He spoke of interest rate adjustments and of "monitoring" acquisitors.

"Monitoring" is a favorite word of the acquisitors themselves for any purpose never explained. It was a favorite word of James Ling's (see p. 159) and of Harold Geneen's (see p. 138) both.

"Volker argues regulation by itself will not work," *The New York Times,* October 20, 2009, reported. Former Federal Reserve Chairman, Paul A. Volker, *did* urge protection of the "invisible hand" and separation of conflicting banking systems (as the repealed Glass-Steagall Act had provided). Giants in pursuit of profits will get in trouble again, he said.

But the repealed limit on the size of banks, enacted during the Great Depression, is closer in time to the Civil War than to our modern age, says Jaime Dimon, CEO of JPMorgan Chase (see p. 5). Though too-titanics-to-let-sink became too titanic because their real business is to buy other businesses, he implores that we not dismantle them. We must keep them big because our European competitors keep their titanics big, he and other CEOs advise.

Where they got that idea they don't say. Not from the Europeans. "The European Commission in Brussels is pressing the UK to shrink its massive banks," *The Wall Street Journal,* November 3, 2009, reported. European ministers "are calling for smaller banks to increase competition." The ministers seek to "eliminate the threat posed by banks so large that they must be rescued by taxpayers no matter how they have conducted their business."

Do those American CEOs argue that the costs of floating their titanics—only beginning with the bailout—do not warrant dismantling them? Are costs of floating them less than the costs averted by our dismantling Standard Oil in 1911? We asserted our right to an open market structure then. And threw it to the ground that awful August day.

Banks that rely on government funds "walk as zombies."[4] How do taxpayers walk who pay that trillion-dollar cost yet never ask that too-titanics be broken apart to restore competitive structure? That structure would have spared them the catastrophe.

NOTES

Frequently Quoted Sources in Order of Reference:

Economic Report on Corporate Mergers, Staff Report, Federal Trade Commission, 1969.

Economic Concentration, Hearings before the Subcommittee on Antitrust and Monopoly of the Committee on the Judiciary, United States Senate, 91st Cong., 2d sess., 1969 and 1970.

Report of the Attorney General's National Committee to Study the Antitrust Laws, 1955.

Brandeis, Louis D. *Other People's Money and How the Bankers Use It* (New York: Harper and Row, 1967 ([1913]).

Investigation of Conglomerate Corporations, Hearings before the Antitrust Subcommittee of the Committee on the Judiciary, House of Representatives, 91st Cong., 2d sess., 1969-70.

Investigation of Conglomerate Corporations, Report by the Staff of the Antitrust Subcommittee of the Committee on the Judiciary, House of Representatives, 92d Cong., 1st sess., June 1, 1971.

Hearings on Military Posture before the Committee on Armed Services, House of Representatives, 92d Cong., 2d sess., April 17, 1972.

Controls over Shipyard Costs and Procurement Practices of Litton Industries, Inc., Pascagoula, Mississippi, Report to the Joint Economic Committee, Congress of the United States, Department of the

Navy. By the Comptroller General of the United States, March 23, 1972.

The Penn Central Failure and the Role of Financial Institutions, Staff Report of the Committee on Banking and Currency, House of Representatives, 92d Cong., 1st sess., January 3, 1972.

Investigation into the Management of the Business of the Penn Central Transportation Co. and Affiliated Companies, Interstate Commerce Commission, Docket No. 35291, March 8, 1972, p. 112.

Emergency Rail Services Legislation, Hearings before the Committee on Interstate and Foreign Commerce and the Subcommittee on Transportation and Aeronautics, House of Representatives, 91st Cong., 2d sess., June, July, and December 1970.

Penn Central Transportation Company: Adequacy of Investor Protection, Hearings before the Special Subcommittee on Investigations of the Committee on Interstate and Foreign Commerce, House of Representatives, 91st Cong., 2d sess., September 24, 1970.

Inadequacies of Protections for Investors in Penn Central and Other ICC-Regulated Companies, Staff Study for the Special Subcommittee on Investigations of the Committee on Interstate and Foreign Commerce, House of Representatives, 92d Cong., 1971.

Hearings before the Committee on the Judiciary on Nomination of Richard G. Kleindienst to be Attorney General, United States Senate, 92d Cong., 2d sess., March 1972.

Part I: The Forces of Concentration

1. Icebergs Unseen

1. *The New York Times,* January 17, 2009.
2. Viral Acharya and Matthew Richardson, *Restoring Financial Stability: How to Repair a Failed System* (NYU-Stern: Wiley Finance, 2009), p. 368.
3. *The Washington Post,* November 8, 2009.
4. Ibid., November 13, 2009.

2. The Offensive Intensifies

1. *Economic Report*, p. 3.
2. Ibid., p. 4.
3. *Dun's Review*, January 1972, p. 31.
4. Senate Select Committee on Small Business, Hearings, March 8, 1973.
5. Ibid.
6. *Economic Report*, p. 51.
7. *Economic Concentration Hearings*, p. 4760.
8. *Economic Report,* p. 111.
9. Ibid., p. 30.

3. Mounting the Counterattack

1. *Economic Concentration Hearings*, p. 5123.
2. Ibid., p. 5177.
3. Ibid., p. 5180.
4. *Attorney General's Study of the Antitrust Laws,* p. 2.

5. Transcript of Joint Economic Committee Hearings, February 21, 1973.

6. "White House Task Force Report on Antitrust Policy," The *Congressional Record*, May 27, 1969, p. S 5643.

7. Senate Report No. 1326, 62d Cong., 3d sess., p. 17.

8. *Other People's Money*, p. 152.

9. *Attorney General's Study of the Antitrust Laws*, p. 117.

10. *Economic Concentration Hearings*, p. 5123.

11. Senate Report No. 1775, 81st Cong., 2d sess., p. 3.

12. House Report No. 1191, 81st Cong., 1st sess., p. 2.

13. Economic Concentration Hearings, p. 5123.

14. Ibid., p. 5177.

15. *Economic Report*, p. 35.

16. *Economic Concentration Hearings*, p. 5122.

17. Ibid.

18. Ibid.

19. Ibid.

20. *Economic Report,* p. 462.

21. Civil Action No. 13320, U.S. District Court, District of Connecticut, filed August 1, 1969.

22. *Hearings on Conglomerate Corporations*, Part 7, p. 1.

23. Ibid., p. 15.

24. Ibid., p. 7.

25. Ibid.

26. *The Congressional Record*, June 16, 1969, p. S 6475.

27. Ibid.

28. Ibid.

29. Ibid., p. 6476.

30. Ibid., p. 6479.

31. Ibid., p. 6478.

32. Ibid.

33. Ibid., p. 6471.

Part II: How to Build an Acquisitor

1. *Hearings on Conglomerate Corporations*, Part 1, p. 150.

4. *Make Friends with a Bank*

1. *Fortune,* March 1968, p. 124.
2. *G &W 1971* [a Gulf & Western publication], p. 17.
3. 12 U.S.C., Sec. 24.
4. *Other People's Money,* p. 103.
5. *Hearings on Conglomerate Corporations*, Part 1, p. 743.
6. Ibid., p. 468.
7. Ibid., p. 338.
8. Ibid., p. 340.
9. Ibid., p. 339.
10. Ibid., p. 37.
11. Ibid., p. 361.
12. Ibid., p. 43.
13. Ibid., p. 47.
14. Ibid., p. 481.
15. Ibid., p. 123.
16. Ibid., p. 173.
17. Ibid., p. 71.
18. Ibid., p. 77.
19. Ibid., p. 774.
20. Ibid., p. 74.
21. Ibid., p. 449.
22. *G &W, 1971*, p. 32.
23. 68 S. Ct. 915 (1948).
24. *Hearings on Conglomerate Corporations*, Part 1, p. 467.
25. Ibid., p. 149.
26. Ibid., p. 147.

27. *G &W 1971,* p. 27.

28. *Hearings on Conglomerate Corporations*, Part 1, p. 103.

29. Ibid., p. 737.

30. Ibid., p. 738.

31. Ibid., p. 454.

32. Ibid., p. 81.

33. Ibid., p. 124.

34. Ibid., p. 84.

35. Ibid., p. 455.

36. Ibid., p. 456.

37. *Hearings on Conglomerate Corporations,* Part 1, p. 125.

38. *Conglomerate Investigation Report,* p. 205.

39. Ibid., p. 204.

40. Ibid.

41. *Hearings on Conglomerate Corporations*, Part 1, p. 474.

42. Ibid., p. 472.

43. Ibid.

44. Ibid., p. 126.

45. *Conglomerate Investigation Report,* p. 198.

46. *Hearings on Conglomerate Corporations*, Part 1, p. 474.

5. Shift the Purchase Debt to the Purchased Company

1. *Hearings on Conglomerate Corporations,* Part 6, p. 66.

2. Ibid., p. 416.

3. Ibid., p. 418.

4. Ibid., p. 575.

5. Ibid., p. 576.

6. Ibid., p. 426.

7. Ibid., p. 156.

8. Ibid., p. 421.

9. Ibid., p. 158.

10. Ibid., p. 431.

11. Ibid., p. 736.

12. LTV 1967 Annual Report.

13. Letter of C. Skeen, LTV president, to George Griffin, vice president for finance, January 11, 1967. House Judiciary Committee files, unpublished.

14. *Hearings on Conglomerate Corporations,* Part 6, p. 574.

15. House Judiciary Committee files, unpublished.

16. *Hearings on Conglomerate Corporations*, Part 6, p. 434.

17. Ibid., p. 162.

18. LTV 1967 Annual Report.

19. LTV 1965 Annual Report.

20. *Hearings on Conglomerate Corporations,* Part 6, p. 736.

21. Ibid., p. 13.

6. *Acquire One Company with the Treasury of Another*

1. *Hearings on Conglomerate Corporations,* Part 4, p. 25.

2. Ibid., p. 12.

3. Ibid., p. 11.

4. Ibid., p. 12.

5. Ibid., p. 22.

6. Ibid., Part 4, p. 25; Part 2, p. 244.

7. Ibid., Part 2, p. 181.

8. Ibid., p. 805.

9. Ibid., p. 179.

10. Ibid., p. 244.

11. Ibid., p. 77.

12. Ibid., p. 181.

13. Ibid., Part 4, p. 45

14. Ibid., p. 60.
15. 12 Code of Federal Regulations 220.
16. *Hearings on Conglomerate Corporations*, Part 4, p. 65.
17. Ibid., Part 2, p. 875.
18. Ibid., p. 880.
19. Ibid., p. 882.
20. *Conglomerate Investigation Report,* p. 229.
21. *Hearings on Conglomerate Corporations,* Part 7, p. 98.
22. Ibid., Part 4, p. 299.
23. Ibid.
24. Ibid., p. 87.
25. Ibid., p. 380.
26. Ibid.
27. Ibid., p. 33.
28. Ibid., p. 349.
29. Ibid.
30. Ibid., p. 31.
31. Ibid., p. 33.
32. Ibid., p. 27.
33. Ibid., p. 39.
34. Ibid., Part 2, p. 224.
35. Ibid., p. 44.
36. Ibid., p. 46.
37. Ibid., p. 48.
38. Ibid., p. 54.
39. Ibid., p. 55.
40. Ibid.
41. Ibid., p. 56.
42. Ibid., p. 75.
43. Ibid., p. 288.

44. Ibid., p. 293.
45. Ibid., p. 195.
46. Ibid., p. 126.
47. Ibid., p. 6.

7. *Accept Delivery from the Pentagon*

1. *Hearings on Conglomerate Corporations,* Part 6, p. 480.
2. Ibid., p. 482.
3. Ibid., p. 484.
4. Ibid., p. 491.
5. Ibid., p. 490.
6. Ibid., p. 173.
7. Ibid., p. 493.
8. Ibid., p. 179.
9. Ibid., p. 508.
10. Ibid., p. 509.
11. Ibid., p. 510.
12. Ibid., p. 183.
13. Ibid., p. 518.
14. Ibid., p. 521.

Part III: The Purpose of the Conglomerate Headquarters

1. *Economic Concentration Hearings*, p. 4584.
2. *Conglomerate Investigation Report,* p. 408.
3. Ibid., p. 411.
4. *Armed Services Hearings*, p. 10569.
5. Ibid., p. 10593.

8. The Purpose of International Telephone & Telegraph

1. *Hearings on Conglomerate Corporations*, Part 3, p. 249.
2. Ibid., p. 161.
3. Ibid.
4. Ibid., p. 158.
5. Ibid., p. 158.
6. Ibid., p. 16.
7. Ibid., p. 22.
8. Ibid.
9. Ibid., p. 23.
10. Ibid.
11. Ibid., p. 94.
12. Ibid., p. 71.
13. Ibid., p. 405.
14. Ibid., p. 101.
15. Ibid.
16. Ibid., p. 10.
17. Ibid., p. 251.
18. Ibid., p. 113.
19. *U. S. v. Penick & Ford, Ltd. Inc.*, 242 F. Supp. 518; *U. S. v. Northwest Inds. Inc.*,
301 F. Supp. 1066.
20. *U. S. v. Internatl. Telephone & Tel. Corp.*, 306 F. Supp. 766.
21. *Hearings on Conglomerate Corporations*, Part 3, p. 481.
22. Ibid., p. 496.
23. Ibid., p. 502.
24. Ibid., p. 121.
25. Ibid., p. 515.
26. Ibid., p. 523.
27. Ibid., p. 529.

28. Ibid., p. 129.

29. Ibid., p. 585.

30. Ibid., p. 787.

31. Ibid., p. 804.

32. Ibid., p. 786.

33. Ibid., p. 787.

34. Ibid., p. 804.

35. Ibid., p. 213.

36. Ibid., p. 215.

37. Ibid., pp. 215, 217.

38. Ibid., p. 216.

39. Ibid., p. 215.

40. Ibid., p. 735.

41. Ibid., p. 217.

42. Ibid., p. 219.

43. Ibid., p. 879.

44. Ibid., p. 225.

45. Ibid., p. 219.

46. Ibid., p. 198.

47. Ibid., p. 196.

48. Ibid., p. 203.

49. Ibid., p. 200.

9. The Purpose of LTV, Inc.

1. *Hearings on Conglomerate Corporations,* Part 6, p. 261.

2. Ibid., p. 90.

3. Ibid., p. 279.

4. Ibid., p. 683.

5. Ibid., p. 261.

6. Ibid., p. 262.

7. Ibid., p. 130.
8. Ibid., p. 131.
9. Ibid., p. 511.
10. Ibid., p. 216.
11. Ibid., p. 83.
12. Ibid., p. 153.
13. Ibid.
14. Ibid., p. 144.
15. Ibid.
16. Ibid., p. 6.
17. Ibid., p. 73.

10. The Purpose of Litton Industries, Inc.

1. *Hearings on Conglomerate Corporations*, Part 5, p. 43.
2. Litton 1970 and 1971 Litton Annual Reports.
3. *Hearings on Conglomerate Corporations*, Part 5, p. 60.
4. Ibid., p. 998.
5. Ibid., p. 1012.
6. Ibid.
7. Ibid., p. 46.
8. *The Wall Street Journal,* March 2, 1973.
9. Litton 1970 Annual Report, p. 18.
10. *The Congressional Record,* April 13, 1972, p. S 6124.
11. *Hearings on Conglomerate Corporations*, Part 5, p. 60.
12. Litton 1970 Annual Report, p. 19.
13. *Armed Services Hearings*, p. 10547.
14. Ibid., p. 10550.
15. Press release of June 16, 1972.
16. *Forbes,* December 15, 1971, p. 18.
17. Ibid.

18. *Armed Services Hearings*, p. 10572.

19. Ibid., p. 10603.

20. Ibid., p. 10658.

21. Ibid.

22. Ibid., p. 10653.

23. Ibid., p. 10665.

24. Ibid., p. 10666.

25. Ibid., p. 10669.

26. Hearings on Conglomerate Corporations, Part 5, p. 61.

27. Armed Services Hearings, p. 10635.

28. Ibid., p. 10608.

29. Ibid., p. 10610.

30. *The Wall Street Journal*, March 2, 1973.

31. *Armed Services Hearings*, p. 10659.

32. *The Journal of Commerce,* March 4, 1969, p. 5.

33. *Armed Services Hearings*, p. 10636.

34. *Hearings on Conglomerate Corporations*, Part 5, p. 61.

35. *Armed Services Hearings*, p. 10618.

36. Ibid.

37. Ibid., p. 10572.

38. Ibid., p. 10574.

39. Ibid., p. 10612.

40. *The Washington Post,* December 23, 1972.

41. *The Wall Street Journal*, December 19, 1972.

42. *Hearings on Conglomerate Corporations*, Part 5, p. 62.

43. Ibid., p. 60.

44. *Report on Controls over Shipyard Costs*, p. 16.

45. Ibid., p. 17.

46. Ibid., p. 4.

47. Ibid., p. 5.

48. Ibid., p. 11.

49. Ibid., p. 13.
50. Ibid., p. 17.
51. Ibid., p. 18.

Part IV: The Counterattack in Disarray

11. The Interstate Commerce Commission

1. *Fortune,* August 1970, p. 106.
2. *The Penn Central Failure*, p. 55.
3. *Fortune,* August 1970, p. 107.
4. *Business Week,* June 27, 1970, p. 98.
5. *The Penn Central Failure*, p. 72.
6. Ibid., p. 33.
7. Ibid., p. 67.
8. Ibid., p. 61.
9. Ibid., pp,. 55, 71.
10. Ibid., p. 25.
11. *Fortune,* August 1970, p. 165.
12. *The Penn Central Failure*, p. 77.
13. *The New York Times,* January 4, 1972.
14. *Fortune,* August 1970, p. 164.
15. *The Penn Central Failure*, p. 273.
16. *Investigation Into the Management of the Business of the Penn Central,* p. 112.
17. *Emergency Rail Services Legislation Hearings*, p. 711.
18. *The Penn Central Failure*, p. 12.
19. George P. Shultz, PBS *Newshour,* December 15, 2009.
20. *Adequacy of Investor Protection Hearing*, p. 5.
21. House of Representatives Report No. 85, 73d Cong., 1st sess., 1933, p. 3.

22. *Adequacy of Investor Protection Hearing,* p. 137.

23. Ibid., p. 112.

24. *Inadequacies of Protections for Investors in Penn Central,* p. 9.

25. *Adequacy of Investor Protection Hearing,* p. 112.

26. *Inadequacies of Protections for Investors in Penn Central,* p. 10.

27. *Adequacy of Investor Protection Hearing,* p. 113.

28. *Inadequacies of Protections for Investors in Penn Central,* p. 25.

29. *Emergency Rail Services Legislation Hearings,* p. 192.

30. *The Penn Central Failure,* p. xi.

31. *Hearing on H.R. 2191 and H.R. 5220 before the House Committee on Interstate and Foreign Commerce,* 76th Cong., 1st sess., 1939.

32. *Fortune,* August 1970, p. 165.

33. *Adequacy of Investor Protection Hearing,* p. 29.

34. Ibid., p. 35.

35. *Inadequacies of Protections for Investors in Penn Central,* p. 36.

36. Ibid., p. 37.

37. Ibid., p. 39.

38. Ibid., p. 40.

39. *Other People's Money,* p. 134.

40. *Investigation into the Management of the Business of the Penn Central,* p. 112.

41. Letter of July 6, 1972, to Morris Goldman.

42. *The New England Investigation,* 27 ICC 560 (1913).

43. *Emergency Rail Services Legislation Hearings,* p. 949.

44. Address of Kenneth H. Tuggle before the American Bar Association, Sections of Public Utility Law and Administrative Law, August 15, 1972.

45. *Emergency Rail Services Legislation Hearings,* p. 952.

46. Commissioner Tuggle's address of August 15, 1972.

47. Emergency Rail Services Legislation Hearings, p. 1041.

48. Commissioner Tuggle's address of August 15, 1972.

49. Emergency Rail Services Legislation Hearings, p. 969.

50. Ibid.

51. Commissioner Tuggle's address of August 15, 1972.

52. *Forbes*, November 15, 1972, p. 29.

53. Ibid.

54. Commissioner Tuggle's address of August 15, 1972.

55. *Emergency Rail Services Legislation Hearings*, p. 1048.

56. Commissioner Tuggle's address of August 15, 1972.

57. *The American Railroad Industry: A Prospectus* (Washington, D. C.: America's Sound Transportation Review Organization, June 1970), p. 21.

12. The Securities and Exchange Commission

1. *Hearings on Conglomerate Corporations*, Part 7, p. 88.

2. Ibid.

3. Ibid., p. 89.

4. Ibid., p. 92.

5. Ibid.

6. *The Work of the Securities and Exchange Commission,* November 1971.

7. *Hearings on Conglomerate Corporations*, Part 7, p. 92.

8. Ibid., p. 93.

9. Ibid., p. 96.

10. Ibid., p. 91.

11. Note I to Financial Statement, LTV 1969 Annual Report, p. 27.

12. *Hearings on Conglomerate Corporations*, Part 3, p. 526.

13. Ibid., p. 358.

14. Ibid., p. 513.

15. Ibid., p. 616.

16. *Conglomerate Investigation Report*, p. 139.

17. Ibid., p. 124.
18. Ibid.
19. Ibid., p. 139.
20. *Hearings before the House Committee on Appropriations for 1971*, 91st Cong., 2d sess., Part 2, p. 1092.
21. Ibid.
22. Ibid., p. 1114.
23. *Hearings before the House Committee on Appropriations for 1973*, 92nd Cong., 2d sess., Part 1, p. 164.
24. Joint Economic Committee files, unpublished.
25. Ibid.
26. Letter of July 14, 1972, ibid.
27. *The Washington Post*, December 21, 1972.
28. Ibid.
29. Ibid.
30. Transcript of Joint Economic Committee Hearings, December 20, 1972, p. 37.
31. *The New York Times*, December 20, 1972.
32. *The Evening Star*, Washington, D. C., December 15, 1972.
33. Ibid.
34. Ibid., December 23, 1972.
35. Transcript of Joint Economic Committee Hearings, December 19, 1972, p. 16.
36. *The New York Times*, December 23, 1972.
37. *The Washington Post*, December 23, 1972.
38. *The Wall Street Journal*, March 16, 1972.
39. *The New York Times*, June 21, 1972.
40. Ibid.
41. Memorandum of Law, House Commerce Committee, Investigations Subcommittee, October 12, 1972, p. 8.
42. Ibid.

43. Transcript of House Commerce Committee Hearings of December 15, 1972, p. 108.
44. Joint press release of October 12, 1972.

13. The Internal Revenue Service

1. *Hearings on Conglomerate Corporations*, Part 7, p. 261.
2. Letter of October 25, 1967, House Judiciary Committee files, unpublished.
3. *Hearings on Conglomerate Corporations,* Part 5, p. 158.
4. Ibid.
5. Ibid., Part 7, p. 97.
6. By letter, House Judiciary Committee files, unpublished.
7. Letter to the Federal Trade Commission, House Judiciary Committee files, unpublished.
8. Title 26 Code of Federal Regulations Sec. 1.368.
9. *The Wall Street Journal,* October 12, 1972.
10. *Kleindienst Nomination Hearings*, p. 103.
11. *The Wall Street Journal,* October 12, 1972.
12. Ibid.
13. Letter of September 25, 1972, from IRS Assistant Commissioner Peter P. Weidenbruch, Jr., to Prof. Boris Bittker.
14. Letter of May 5, 1972, from IRS Acting Commissioner Raymond Harless to Ralph Nader, Alan Morrison, and Reuben Robertson.

14. The Department of Justice

1. Plaintiffs brief, *Nader, et al. v. Cotter, et al.*, No. 166205, Connecticut Superior Court, Hartford County, January 27, 1972, p. 23.

2. Ibid., p. 31.

3. Ibid., p. 12.

4. *Kleindienst Nomination Hearings*, p. 118.

5. *TheWashington Post,* March 25, 1972.

6. *Kleindienst Nomination Hearings*, p. 118.

7. Ibid., p. 265.

8. Ibid., p. 268.

9. Ibid., p. 265.

10. Ibid., p. 268.

11. ITT 1972 1st Quarter Report.

12. *Kleindienst Nomination Hearings,* p. 266.

13. Ibid., p. 267.

14. Ibid., p. 289.

15. Ibid., p. 328.

16. Ibid.

17. Ibid., p. 110.

18. Ibid., p. 644.

19. Ibid., p. 104.

20. Ibid., p. 109.

21. Ibid., p. 114.

22. Ibid., p. 111.

23. ITT 1971 Annual Report.

24. *Hearings on Conglomerate Corporations*, Part 3, p. 21.

25. ITT 1972 1st Quarter Report.

Conclusion
Icebergs Dead Ahead

1. Address of Richard W. McLaren before the Federal Bar Association, Council on Antitrust and Trade Regulation, September 17, 1970.

2. Address of Thomas E. Kauper before the American Bar Association, Section of Corporate, Banking and Business Law, October 27, 1972.
3. *The Wall Street Journal*, April 24, 1972.
4. Transcript of Senate Foreign Relations Committee Hearings, March 22, 1973, p. 47.
5. Ibid., March 21, 1973, p. 23.
6. *Conglomerate Investigation Report*, p. 439.

Index